Hopmann / Greif / Wolters
Training in Plastics Technology

Christian Hopmann
Helmut Greif
Leo Wolters

Training in Plastics Technology

3rd Edition

HANSER

Print-ISBN: 978-1-56990-910-2
E-Book-ISBN: 978-1-56990-935-5

All information, procedures, and illustrations contained in this work have been compiled to the best of our knowledge and is believed to be true and accurate at the time of going to press. Nevertheless, errors and omissions are possible. Neither the authors, editors, nor publisher assume any responsibility for possible consequences of such errors or omissions. The information contained in this work is not associated with any obligation or guarantee of any kind. The authors, editors, and publisher accept no responsibility and do not assume any liability, consequential or otherwise, arising in any way from the use of this information – or any part thereof. Neither do the authors, editors, and publisher guarantee that the described processes, etc., are free of third party intellectual property rights. The reproduction of common names, trade names, product names, etc., in this work, even without special identification, does not justify the assumption that such names are to be considered free in the sense of trademark and brand protection legislation and may therefore be used by anyone.

The final determination of the suitability of any information for the use contemplated for a given application remains the sole responsibility of the user.

Bibliographic information of the German National Library:
The German National Library lists this publication in the German National Bibliography; detailed bibliographic data are available on the Internet at *http://dnb.d-nb.de*.

This work is protected by copyright.
All rights, including those of translation, reprint, and reproduction of the work, or parts thereof, are reserved. No part of this work may be reproduced in any form (photocopy, microfilm, or any other process) or processed, duplicated, transmitted, or distributed using electronic systems, even for the purpose of teaching – with the exception of the special cases mentioned in §§ 53, 54 UrhG (German Copyright Law) – without the written consent of the publisher.

© 2024 Carl Hanser Verlag GmbH & Co. KG, Munich
www.hanserpublications.com
www.hanser-fachbuch.de
Editor: Mark Smith
Production Management: Cornelia Speckmaier
Coverconzept: Marc Müller-Bremer, *www.rebranding.de,* Munich
Cover design: Max Kostopoulos
Cover picture: © Carl Hanser Verlag, unter Verwendung von Grafiken von © shutterstock.com/Eywa
Typesetting: Eberl & Koesel Studio, Kempten
Printed and bound by: Hubert & Co – eine Marke der Esser bookSolutions GmbH, Göttingen
Printed in Germany

Preface to the Third Edition

We are pleased that you have decided to study this book, which is available in both print and e-book formats.

The basis of the book was created almost 50 years ago as part of a research project lasting several years with the aim of finding and developing suitable methods of knowledge transfer for plastics technology. In 1976, a first German edition was published under the title *Lernprogramm Technologie der Kunststoffe* ("Training Program Technology of Plastics"), which was developed by the Institute of Plastics Processing (IKV) at RWTH Aachen University in cooperation with the Institute of Educational Science at RWTH Aachen University.

The editors were Prof. Georg Menges (Head of the Institute of Plastics Processing at RWTH Aachen University), Prof. Johannes Zielinski (Director of the Institute for Educational Science at RWTH Aachen University) and Ulrich Porath, a research engineer at the IKV.

The preface to the first edition in 1976 started with the statement:

"It is impossible to imagine our daily lives without plastics. We take this material in our hands as a matter of course without having dealt with it in any detail …"

Today, 47 years later, this statement is more valid than ever, as plastic materials have opened application ranges in almost all areas of life and will continue to do so in the future.

This comprehensively revised, new English edition of the text- and workbook still pursues the same goal of introducing the reader to the world of plastics and imparting the essential basics about the material and its machining and processing. It has been revised professionally, technically, and pedagogically, and further lessons on new topics have been added.

The book is a collaborative effort. The authors wish to thank all those who have contributed to its success – especially the companies and institutions that have provided ample visual material or information.

At this point, we would like to express our sincere thanks to all those who have contributed to the revisions of the different editions: Dr. Johann Thim, Hans Kaufmann, Prof. Walter Michaeli as well as Franz-Josef Vosseburger.

We hope you enjoy learning and working with this new edition.

Christian Hopmann

Helmut M. Greif

Leo Wolters

We highly appreciate any suggestions for improvements (office@ikv.rwth-aachen.de).

Editor's Preface to the Second Edition

This short volume is intended as a text and workbook for technicians employed in the plastics industry. The original German language edition was prepared at the behest of the government of the Federal Republic of Germany, the German Federal Association of Employers in the Chemical Industry and the German Chemical Workers Union. The book was put together and written at the Institut für Kunststoffverarbeitung (IKV) (Institute for Plastics Processing) at the Technical University of Aachen by Prof. Walter Michaeli with Helmut Greif, Hans Kaufmann and Franz-Josef Vosseburger.

Germany has long realized the necessity of organized post-high school teaching of trades to aspiring technicians. It is envisaged as the modem equivalent of the medieval apprenticeship program. Plastics processing is recognized as one of the modem trades deserving this treatment. The German program is described in some detail in Appendix 1.[1] It is one of known success as witnessed by the success of Germany's many manufacturing firms. The skill of their well-trained technicians is proverbial.

In the United States, there has been increased interest in recent years in the post-high school training of technicians. However, little attention has been given to the system long established in Germany which could serve as a model.

It is hoped that this book will find use in educational programs for technicians taught both within industry and in schools/programs dedicated to this purpose.

James L. White

Institute of Polymer Engineering

University of Akron

---

[1] This refers to Lesson 21 in the 3rd edition.

The Authors

Professor Dr.-Ing. Christian Hopmann

Prof. Christian Hopmann holds the Chair for Plastics Processing and is director of the IKV – Institute for Plastics Processing in Industry and Craft at RWTH Aachen University in Germany. He is also co-founder of the AZL – Aachen Center for Lightweight Production and Vice Dean of the faculty for Mechanical Engineering of RWTH Aachen University. His interests lie in fundamental and applied research in plastics technology with a particular focus on digitization and simulation, lightweight technologies, and the circular economy. Hopmann is the principal investigator and member of the steering committee of the Federal Cluster of Excellence "Internet of Production". He initiated the Polymer Innovation Center 4.0, which addresses the domain specific realization and implementation of digitization in the plastics industry, with a particular focus on SME.

Hopmann received the Innovation Award of the Federal German state of North Rhine-Westphalia in 2014. He was appointed visiting professor at the Beijing University of Chemical Technology, Beijing/China in 2017 and fellow of the Society of Plastics Engineers (SPE), CT/USA, in 2019. Hopmann has served as international representative of the Polymer Processing Society (PPS) since 2021 and has been a member of the board of directors and the scientific advisory board as well as chairman of the board of the material engineering division of the VDI – The Association of German Engineers – since 2022.

Dr. Dipl.-Ing. Helmut Greif M. A.

Until 2016, Dr. Helmut Greif was Managing Director of the Aachen Society for Innovation and Technology Transfer (AGIT). He studied mechanical engineering, with a focus on construction technology, at the Aachen University of Applied Sciences, and sociology/political science/education at RWTH Aachen University, where he also received his interdisciplinary doctorate under Prof. Hörning (sociology) and Prof. Michaeli (mechanical engineering). After working in industry and business as a factory planner (company: agiplan) and then as managing director of a training and qualification institution (Dr. Reinhold Hagen Foundation), he was head of the HPI (Heinz Piest Institute) at the University of Hanover in the faculty of mechanical engineering before moving to AGIT in Aachen in 2007.

Dipl.-Ing. Leo Wolters

Leo Wolters was Managing Director of the Training and Further Education Department at the Institute for Plastics Processing (IKV) in Industry and Craft at RWTH Aachen University from 1995 to 2021. He studied mechanical engineering at the Cologne University of Applied Sciences, majoring in production engineering. Since September 1984, he has been engaged in technology transfer and training and further education in the field of plastics processing at the IKV in Aachen and has been actively involved in numerous national and international committees as well as standards committees.

■ Translation

Dr.-Ing. Marion Hopmann

Marion Hopmann studied mechanical engineering with particular focus on plastics processing at RWTH Aachen University. She received her doctoral degree at the Institute for Plastics Processing (IKV) in Industry and Craft at RWTH Aachen University with a thesis on additive manufacturing. After a career in the automotive industry at Ford Motor Company and Visteon she gave lectures on injection molding at FH Aachen University of Applied Sciences in Jülich and is currently working as consultant.

Notes
How to Use this Book

Introduction

This text and workbook provide an introduction to the world of plastics. The use of the plural "plastics" instead of the singular "plastic" indicates that we are dealing with a variety of different materials that can differ significantly in processibility or in their response to the influence of heat. Nevertheless, they are all classified as plastics as they are all synthetically produced, meaning newly composed, and thus do not occur in this form in nature.

Lessons

"Training in Plastics Technology" is divided into educational units that can be described as lessons, with each lesson covering a distinct subject area. The lessons, approximately equal in length, are designed so that they can be arranged by each student in a meaningful educational sequence.

Key Questions

The key questions at the beginning of each lesson are intended to help the student approach the subject matter with certain questions in mind. The answers to these questions will become clear after the student has worked through the lesson.

Prerequisite Knowledge

It is not necessary to study the lessons in any sequence. Information is provided in each lesson that indicates which lessons or content are important for understanding the lesson at hand.

Subject Area

The lessons can each be assigned to superordinate subject areas. Each lesson starts with a note indicating the subject area to which it belongs.

Performance Review

The review questions at the end of each chapter serve to verify the acquired knowledge. The correct answer must be selected from the list of answers provided and entered in the space provided in the text. The answers can be checked from the solutions at the end of the book. If the selected answer is incorrect, the corresponding topic should be worked through again.

Example: "Optical Data Carrier" (CD, CD-ROM, DVD, Blu-ray Disc)

To increase the understanding of plastics and improve thinking in contexts, a molded part made of plastic has been chosen to serve as an example, and will be referred to in many of the lessons in the book. This product is used to show why, for example, a particular plastic is ideal for manufacturing "optical data carriers", such as a CD, and to ask whether this plastic can be recycled.

Additional Information

Literature, glossary, professions, and abbreviations: The appendixes provide supplementary material on plastics for the interested reader. The selected bibliography can help with finding information on further technical literature. The glossary is intended to contribute to a standardized understanding of the terms used, and it can serve as a kind of short encyclopedia. The information on the job description of "Process Technician for Plastics and Rubber Engineering" and "Materials Tester Focus: Plastics" offers an opportunity to find out more about the tasks of these German plastics professions and their different specializations as well as about further training opportunities and promotion prospects in this area. A list of abbreviations, both general and plastics-specific, is provided to facilitate an understanding of the technical content. The most important physical values and their formula symbols are also provided.

Notes
Abbreviations and Symbols

Physical Quantities

Formula Symbol	Explanation	Unit
A	Area	m²
a	Center-to-center distance	m
α	Thread angle (Greek: "alpha")	°
Å	Ångström (1 Å = 10^{-10} m)	m
b	Channel width	m
D	Screw diameter	m
d	Core diameter	m
e	Land width	m
E-modulus	Young's modulus (modulus of elasticity)	MPa
ε	Strain (Greek: "epsilon")	%
F	Force	N
$\dot{\gamma}$	Shear rate (Greek: "gamma dot")	1/s
h	Flight depth	m
η	Viscosity (Greek: "eta")	Pa s
i	Number of flights	n (number)
J	Joule	W s = N m
λ	Heat conductivity (Greek: "lambda")	W/m K
L/D	Screw length/screw diameter ratio	1
M	Torque	N m
Mt	Million tons	
P	Power	W
φ	Screw pitch angle (Greek: "phi")	°
$\dot{Q}$	Volumetric flow rate	m³/s
Q	Heat quantity	J
$qm = \dot{m}$	Mass flow	kg/s
R	Resistance	Ω (Ohm)

Formula Symbol	Explanation	Unit
s_k	Screw clearance	m
σ	Tensile strength (Greek: "sigma")	N/m^2 or Pa
t	Screw pitch	
t	Ton (metric) or tonne	1000 kg
T	Temperature	°C
T	Trillion (10^9)	
τ	Shear stress (Greek: "tau")	N/m^2
T_f	Flow temperature range (FT)	°C
T_g	Glass transition or softening temperature range (GT or ST)	°C
T_c	Crystalline melting temperature range (CM)	°C
v	Velocity	m/s

Plastics

Abbreviation	Explanation
ABS	Acrylonitrile butadiene styrene copolymers (amorphous copolymers)
BR	Polybutadiene (general purpose rubber; butadiene rubber)
C	Carbon (Latin: "carbonium")
CAMPUS	Computer Aided Material Preselection by Uniform Standards
CFRP	Carbon fiber reinforced plastic (carbon fiber composite material (CF) with a polymer matrix)
Cl	Chlorine
CMR	Crystalline melting temperature range (also T_c)
CR	Polychloroprene (specialty type of rubber)
DT	Decomposition temperature
EP	Epoxy resins
EVOH	Ethylene/vinyl alcohol copolymer
EX	Extrusion
F	Fluorine
FIT	Fluid injection technology
FRP	Fiber reinforced polymers
FT	Flow temperature range (also T_f)
GIT	Gas injection molding, also gas-assisted injection molding
GKV	*Gesamtverband Kunststoffverarbeitende Industrie* (German Association of Plastics Converters)
GMT	Glass mat reinforced thermoplastics
GRP	Glass fiber reinforced plastics – composite materials made of glass fibers (GF) and a polymer matrix

Abbreviation	Explanation
H	Hydrogen (Latin: "Hydrogenium")
IKV	*Institut für Kunststoffverarbeitung* (Institute for Plastics Processing)
IM	Injection Molding
MFR	Melt flow rate, is replaced by melt volume rate (MVR)
MVR	Melt volume rate, colloquially also MVI (melt volume index)
N	Nitrogen (Latin: "nitrogenium")
NR	Natural Rubber
O	Oxygen (Latin: "oxygenium")
PA	Polyamide (semicrystalline thermoplastic)
PC	Polycarbonate (amorphous thermoplastic)
PE	Polyethylene (semicrystalline thermoplastic)
PE-HD	High-density polyethylene
PE-LD	Low-density polyethylene
PEEK	Polyether ether ketone (semicrystalline thermoplastic, heat-resistant)
PES	Polyether sulfone (amorphous thermoplastic)
PFT	Polyethylene terephthalate (semicrystalline thermoplastic)
PF	Phenol formaldehyde
PMMA	Polymethyl methacrylate (amorphous thermoplastic)
POM	Polyoxymethylene (semicrystalline thermoplastic), also called polyacetal
PP	Polypropylene (semicrystalline thermoplastic)
PPI	Plastics processing industry
PS	Polystyrene (amorphous/semicrystalline thermoplastic)
PUR	Polyurethane (elastomer)
PVC	Polyvinyl chloride (amorphous thermoplastic)
RIM	Reaction injection molding
SBR	Styrene butadiene rubber
SMC	Sheet molding compound
ST	Softening temperature range (also T_g = glass transition temperature range)
ST	Softening temperature
UP	Unsaturated polyester resin
WIT	Water-assisted injection molding

General Terms

Abbreviation	Explanation
AI	Artificial intelligence
Al	Aluminum
APR	Accident Prevention Regulations (see also VBG)

Abbreviation	Explanation
ASI	Austrian Standards International
AT (AUT)	Austria (.at)
BBiG	*Berufsbildungsgesetz* – (German) vocational training act
BEM	Boundary element method
BG	*Berufsgenossenschaft* – (German) occupational insurance association
BGR	*Berufsgenossenschaftliche Regeln* – (German) occupational insurance association rules
BGV	*Berufsgenossenschaftliche Vorschriften* – (German) regulations of the trade association
BIBB	*Bundesinstitut für Berufsbildung* – (German) Federal Institute for Vocational Training
Blu-ray disc	HD-DVD = High-Density Digital Versatile Disc
CAD	Computer aided design
CAE	Computer aided engineering
CAM	Computer aided manufacturing
CAQ	Computer aided quality
CD	Compact disc
CH (CHE)	Switzerland (.ch)
CIM	Computer integrated manufacturing
CIP	Continuous improvement process
CNC	Computerized numerical control
DE (DEU)	Germany (.de)
DGQ	*Deutsche Gesellschaft für Qualität* – German society for quality
DGUV	*Deutsche Gesetzliche Unfallversicherung* – German social accident insurance
DIN	*Deutsches Institut für Normung* – German equivalent of American National Standards Institute ANSI
DSD	*Duales System Deutschland* – Dual System Germany
DVD	Digital Versatile Disc
EC	Is replaced by EU – European Union
ECS	European Committee for Standardization
EDP	Electronic data processing
EMG	European machine guidelines
EMS	Environmental management system
EN	European Norm
EU	European Union (see also EC)
FEM	Finite elements method
FVM	Finite volume method
GC	German constitution

Abbreviation	Explanation
HD	High density
IMS	Integrated management systems (see also QM, EMS)
IR	Infrared
ISO	International Standard Organization
JIS	Just in sequence
JIT	Just in time
KrWG	*Kreislaufwirtschaftsgesetz* – German circular economy law
LCA	Life cycle assessment
LD	Low density
LIL	Lower intervention limit (see also UIL)
log	Logarithmic (not linear)
LP	Long-playing record
MT	Machine tool
OHS	Occupational health and safety
ÖN	*Österreichische Norm* – Austrian Norm
OS	Operations scheduling
OSHA	Occupational Safety and Health Act
PCDA	Plan-check-do-act
PGE	Planetary gear extruder
QA	Quality assurance (see also IMS)
QM	Quality management (see also IMS)
QMS	Quality management system
QRK	Quality control chart
SF	Substitute fuels
SME	Small and medium-sized enterprises
SN	Swiss Norm
SNV	Swiss Standards Association
SPC	Statistical process control
TQM	Total quality management
UIL	Upper intervention limit (see also LIL)
UV	Ultraviolet
VBG	Regulation of the Trade Association (see also APR)
VDA	Association of the German Automotive Industry
VerpackG	German Packaging Act from 2019 (replaces the German VerpackV)
VerpackV	German Packaging Act from 1998, new regulation 2014)
WIP	Waste incineration plant

Academic Degrees

Abbreviation	Explanation
Dipl.-Ing.	Diplom-Ingenieur (old term in Germany for M. Eng. or M. Sc.)
B. Eng.	Bachelor of Engineering
B. Sc.	Bachelor of Science
M. Eng.	Master of Engineering
M. Sc.	Master of Science

Introduction
Plastic – An Artificial Material?

Key Questions Where do we encounter plastics in everyday life?
How long have we been using plastics?
What is a compact disc (CD) made of?

Contents Plastics – Part of Our Everyday Life
Plastics – Versatile Materials
Plastics – Young Materials

Plastics – Part of Our Everyday Life

In our environment, plastics have become perfectly acceptable as a matter of course in everyday use. People don't think about why these products are made of plastic when they use freezer bags or cell phones.   plastics ...

Why are more and more drinking bottles being made of plastic instead of glass?

Here, weight plays the most crucial role. The lighter, plastic bottle is stable enough to transport the liquid it contains. It is more energy-efficient to manufacture and saves fuel as well as CO_2 because less weight is being transported. The consumer also benefits from carrying a lighter, plastic bottle.   ... are lightweight

Why are power cables coated with plastic and not, for instance, with porcelain or fabric?

Plastic sheathing is more flexible than porcelain and tougher than fabric, yet it insulates the cable just as well, if not better.   ... insulate against electric current and can be flexible

Why is a refrigerator interior lined with plastic?

Because plastic is, on one hand, rugged. On the other, it is a poor conductor of heat, and so the low temperatures can be maintained better. Furthermore, the surfaces are easy to clean.   ... insulate against heat

The opposite is the case, for example, for the insulation of houses. Here, foamed plastics help to keep the heat in the house for much longer. Heating costs, but also CO_2 emissions, are significantly reduced.   ... insulate against cold

Why is a CD made of plastic?

Because the plastic polycarbonate (PC) is as translucent as glass. At the same time, it is much lighter than glass and not as fragile.

... are low-cost materials

Of course, we must also consider the price in all these examples. Using plastics is often the more cost-effective technical solution, especially for mass-produced articles. Why this is so, and which problems are often overlooked in this context (e.g. waste disposal), we will examine later.

Plastics – Multifunctional Materials

wood
natural rubber

Before plastics became known, only nature provided lightweight materials. Wood is easy to process and is strong and flexible. It can also be permanently shaped by special processes. Natural rubber, a raw material for synthetic rubber, is elastic and stretchable.

natural materials

All technical problems cannot be solved with the properties of natural materials, however. This triggered a search for new materials possessing the required properties. Not until the twentieth century did chemists learn enough about the molecular structure of natural materials (e.g., natural rubber) to be able to produce these materials artificially. The heat-insulating neoprene (for wetsuits), which came to market in 1930 and was produced from rubber, was the first major application of this new group of materials.

Lego bricks

Another example illustrating the diversity of plastics is the "Lego bricks" that were launched in 1958. Initially they were made from cellulose acetate and later from ABS. The high quality of this well-known plastic product is apparent from the fact that even after 50 years, the precision fit is still fully guaranteed.

ideal properties

The properties of plastics produced today are often far superior to those of natural materials. For the most diverse purposes, we now have materials whose properties are ideally customized to the intended application.

material properties

It is impossible to determine the purpose for which a plastic article is best suited by observing its external appearance. We also need to know something about the internal structure of the material. It gives us information about density, conductivity, permeability, or solubility, for example. In other words, it tells us the "material-specific properties".

Plastics – Young Materials

plastics model
Nobel Prize

The systematic conversion of natural substances into the materials known today as "plastics" began in the 19th century. However, they did not attain commercial significance until the 1930s, when Prof. Hermann Staudinger developed his model picture of the structure of plastics. A German chemist, Staudinger (1881 to 1965) received the Nobel Prize for this research in 1953.

The global boom in the plastics industry began after World War II. Coal was initially used as the basic material, and it was not until the mid-fifties that the switch was made to petroleum. The advantage of this change was that the previously worthless refining fractions that occurred as separation products in the process of cracking crude oil could be put to effective use. The rapid increase in plastics production experienced a moderate setback during the 1973 oil crisis. Nevertheless, the materials have recorded above-average, dynamic growth up to the present day.

petroleum

World production of plastics shows a continuous growth rate of 3 to 5% per year (see Figure 1).

substituting traditional materials

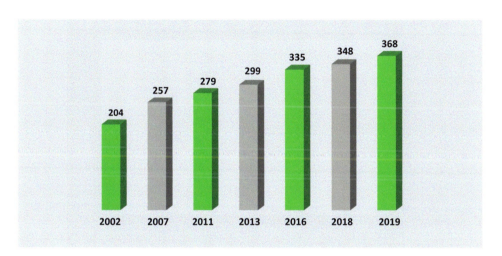

Figure 1 Worldwide production of plastics in millions of tons [based on: Plastics Europe]

However, plastics can only be used with optimal effectiveness when their special characteristics are considered. Particularly when they are substituted for traditional materials such as wood or metal, a design suitable for plastics must be taken into consideration that will allow the plastics to bring their many possibilities to the application. It is important to be familiar with the appropriate processing methods, as well as the corresponding characteristic material values.

Such a plastic-based approach requires a fundamental understanding of the manufacturing and processing methods as well as the material characteristics. This book is intended to provide a first fundamental and comprehensive overview of the subject of plastics. We intend to follow a technical plastic component on its way from the starting material, crude oil, through production, to final disposal through recycling. This part will be a compact disc (CD) or DVD, which should be familiar to everyone. It thus makes an ideal example of modern plastics processing.

compact disc (CD)

Figure 2 shows a CD produced via the injection molding process and its dimensions.

CD

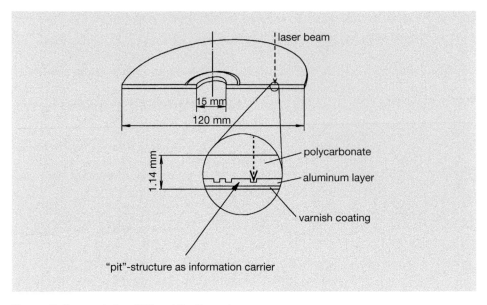

Figure 2 Compact disc (CD) and its dimensions

DVD
Blu-ray (HD-DVD)

The modern siblings of the CD, for example the DVD (Digital Versatile Disc) and the HD-DVD or Blu-ray disc, are not produced using the classic injection molding process. Due to their multilayer structure, significantly thinner discs must be produced. They are injection compression molded and glued together in a further step.

long-playing record (LP)
polyvinyl chloride (PVC)

It is interesting to note in this context that the long-playing record (LP) was a plastic predecessor technology of the CD for high-quality music. The LP came onto the market in 1948. At that time, this new material group of "plastics" contributed decisively to the worldwide success of a recording medium. The LP is made of PVC (polyvinyl chloride), a material that is highly resistant to external influences and is the same material used to produce plastic windows, for example.

Contents

Preface to the Third Edition V

Editor's Preface to the Second Edition VII

The Authors IX
    Translation X

Notes   How to Use this Book XI

Notes   Abbreviations and Symbols XIII

Introduction   Plastic – An Artificial Material? XIX

Lesson 1   Plastics Fundamentals 1
    1.1   What are "Plastics"? 1
    1.2   What are Plastics Made of? 2
    1.3   How to Classify Plastics? 2
    1.4   How are Plastics Identified? 4
    1.5   What are the Physical Properties of Plastics? 5
    1.6   Performance Review – Lesson 1 8

Lesson 2   Raw Materials and Polymer Synthesis 9
    2.1   Raw Materials for Plastics 9
    2.2   Monomers and Polymers 11
    2.3   Polyethylene Synthesis 13
    2.4   Methods of Polymer Synthesis 15
    2.5   Performance Review – Lesson 2 21

Lesson 3	**Classification of Plastics**	23
	3.1 Bonding Forces in Polymers and Their Temperature Behavior	24
	3.2 Identification of Categories of Plastics	26
	3.3 Thermoplastics	26
	3.4 Cross-Linked Plastics (Elastomers and Thermosets)	28
	3.5 Fabrication and Processing Methods	30
	3.6 Methods for Shaping Thermoplastics	31
	3.7 Performance Review – Lesson 3	32
Lesson 4	**Deformation Behavior of Plastics**	35
	4.1 Behavior of Thermoplastics	35
	4.2 Amorphous Thermoplastics	36
	4.3 Semicrystalline Thermoplastics	37
	4.4 Behavior of Cross-Linked Plastics	39
	4.5 Performance Review – Lesson 4	41
Lesson 5	**Time-Dependent Behavior of Plastics**	43
	5.1 Behavior of Plastics Under Load	43
	5.2 Effect of Time on Mechanical Behavior	45
	5.3 Recovery Behavior of Plastics	46
	5.4 Dependence of Plastics on Temperature and Time	47
	5.5 Performance Review – Lesson 5	51
Lesson 6	**Physical Properties**	53
	6.1 Density	53
	6.2 Thermal Conductivity	54
	6.3 Electrical Conductivity	55
	6.4 Transparency	56
	6.5 Material Characteristics of Plastics	57
	6.6 Performance Review – Lesson 6	61
Lesson 7	**Fundamentals of Rheology**	63
	7.1 Fundamentals	63
	7.2 Flow and Viscosity Curves	66

	7.3	Flow Behavior of Plastic Melts	66
	7.4	Melt Flow Index (MFI)	68
	7.5	Performance Review – Lesson 7	70
Lesson 8	**Plastic Applications**		**71**
	8.1	Fundamentals	72
	8.2	Requirements Criteria – Material Selection – Manufacturing Processes	72
	8.3	Examples of Plastic Applications	76
	8.4	Performance Review – Lesson 8	82
Lesson 9	**Plastics Compounding**		**83**
	9.1	Fundamentals	83
	9.2	Metering	85
	9.3	Mixing	85
	9.4	Plasticizing	87
	9.5	Pelletizing	89
	9.6	Crushing	91
	9.7	Performance Review – Lesson 9	92
Lesson 10	**Extrusion**		**93**
	10.1	Fundamentals	93
	10.2	Extrusion Lines	94
	10.3	Coextrusion	103
	10.4	Extrusion Blow Molding	104
	10.5	Blown Film Process	107
	10.6	Performance Review – Lesson 10	108
Lesson 11	**Injection Molding**		**111**
	11.1	Fundamentals	111
	11.2	Injection Molding Machine	113
	11.3	The Injection Mold	117
	11.4	Process Flow	119
	11.5	Other Injection Molding Processes	123
	11.6	Examples and Products	125
	11.7	Performance Review – Lesson 11	126

Lesson 12	**Fiber-Reinforced Composites (FRC)**	**127**
12.1	Fundamentals	127
12.2	Materials	128
12.3	Process Flow	129
12.4	Hand Lay-Up	130
12.5	Automated Processing Methods	131
12.6	Performance Review – Lesson 12	135
Lesson 13	**Plastic Foams**	**137**
13.1	Fundamentals	137
13.2	Properties of Foams	138
13.3	Foam Production	141
13.4	Examples and Products	144
13.5	Performance Review – Lesson 13	145
Lesson 14	**Thermoforming**	**147**
14.1	Fundamentals	147
14.2	Process Steps	149
14.3	Technical Installations	151
14.4	Examples and Products	152
14.5	Performance Review – Lesson 14	153
Lesson 15	**Additive Manufacturing**	**155**
15.1	Fundamentals	155
15.2	Plastics for Additive Manufacturing	158
15.3	Operating Steps and Process Parameters	160
15.4	Examples and Products	163
15.5	Performance Review – Lesson 15	164
Lesson 16	**Plastic Welding**	**165**
16.1	Fundamentals	165
16.2	Process Steps	166
16.3	Welding Processes	167
16.4	Examples and Products	172
16.5	Performance Review – Lesson 16	173

Lesson 17	**Machining Plastics** **175**
	17.1 Fundamentals 175
	17.2 Cutting Processes 176
	17.3 Performance Review – Lesson 17 180

Lesson 18	**Bonding of Plastics** **181**
	18.1 Fundamentals 181
	18.2 Bondability and Adhesive Joints 184
	18.3 Classification of Adhesives 186
	18.4 Bonding Process 187
	18.5 Examples and Products 188
	18.6 Performance Review – Lesson 18 189

Lesson 19	**Plastic Waste** **191**
	19.1 Fundamentals 191
	19.2 Plastics Production and Its Applications . 192
	19.3 Plastic Products and Lifetime 194
	19.4 Avoiding and Recycling Plastic Waste 196
	19.5 Circular Economy in Plastics Business 197
	19.6 Performance Review – Lesson 19 199

Lesson 20	**Plastics Recycling** **201**
	20.1 Fundamentals 201
	20.2 Mechanical Recycling 203
	20.3 Chemical Recycling 206
	20.4 Thermal Recovery 207
	20.5 Recycling of Plastic Waste 209
	20.6 Examples and Products 209
	20.7 Performance Review – Lesson 20 213

Lesson 21	**Qualification in Plastics Processing** **215**
	21.1 Fundamentals 215
	21.2 Plastics Training in Industry 216
	21.3 Plastics Training in the Skilled Trades/Crafts Sector 223

Appendix	Selected Literature	227
Appendix	Glossary of Plastics Technology	233
Appendix	Answers	255

Lesson 1
Plastics Fundamentals

Subject Area	Plastics Fundamentals
Key Questions	How can plastics be defined?
	What are plastics made of?
	How are plastics classified?
	What plastic is a CD made of?
	Are plastics recyclable?
	What properties do plastics have?
	Where can we find plastics in use?
Contents	1.1 What are "Plastics"?
	1.2 What are Plastics Made of?
	1.3 How to Classify Plastics?
	1.4 How are Plastics Identified?
	1.5 What are the Physical Properties of Plastics?
	1.6 Performance Review – Lesson 1

■ 1.1 What are "Plastics"?

The name "plastic" does not stand for a single material. Just as "metal" is used to describe more than just iron or aluminum, the name "plastic" is the generic term for many materials that differ in structure, properties, and composition. The properties of plastics are so diverse that they often replace or supplement conventional materials, such as wood or metal. *generic term*

All plastics have one thing in common, however. They are the result of the tangling or linking of very long molecular chains (chain molecules) known as "macromole- *macromolecules*

cules" (Greek: macros = large). These macromolecules often consist of more than 10,000 individual structural elements. In these molecular chains, the individual building blocks are arranged one after the other like pearls on a necklace. The plastic can be thought of as something similar to a ball of wool made up of many individual threads. It is very difficult to pull a single thread out of the ball. The situation is similar in a plastic, in which the macromolecules "hold on" to each other. Since macromolecules, and thus the plastics, are made up of many individual structural elements, the monomer molecules (Greek: monos = individual, meros = part), they are also generally called polymers (Greek: poly = many).

definition

Plastics are materials whose essential components consist of macromolecular, organic compounds that are created synthetically or by the conversion of natural products. Usually, these materials can be shaped or undergo plastic deformation when processed under certain conditions (e.g., heat or pressure).

■ 1.2 What are Plastics Made of?

monomers

The basic substances for polymers are called "monomers". It is often possible to produce several different polymers from the same individual basic substances by varying the manufacturing process or by creating different mixtures.

raw materials refinery products

The basic materials for monomers are mainly crude oil and natural gas. Since carbon is the only relevant material for production, monomers could theoretically also be produced from wood, coal or even atmospheric CO_2. However, these substances are not yet being used because production from gas and oil is cheaper. Several years ago, some monomers were waste materials in the production of gasoline or fuel oil. Today, the high consumption of plastics makes it necessary to specifically produce these "waste monomers" in refineries.

■ 1.3 How to Classify Plastics?

Three overall groups of plastic materials are distinguished from one another. Figure 1.1 presents each of these groups, along with examples.

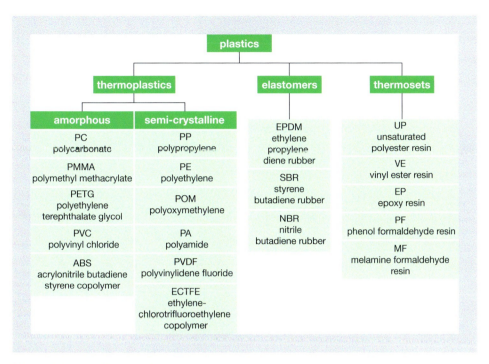

Figure 1.1 Classification of plastics

Thermoplastics (Greek: thermos = warm; plasso = shape, shapable) are fusible and soluble. They can be remelted several times and are soluble or at least swellable in many solvents. They vary from being soft to tough or hard and brittle at room temperature. A distinction is made between amorphous (Greek: amorphos = formless) thermoplastics, which resemble glass with respect to their molecular structure and are crystal clear, and semicrystalline thermoplastics, which have a opaque milky appearance. If a plastic is as transparent as glass, it is reasonably safe to conclude that it is an amorphous thermoplastic. Thermoplastics make up the largest proportion of plastics by quantity.

thermoplastics
amorphous thermoplastics semicrystalline thermoplastics

We will therefore make the cover of the sleeve for our CD from an amorphous material because it is supposed to be transparent in order to be able to read the list of titles. The plastic of the CD itself is also transparent. Usually, it is vapor-coated on one side with aluminum (the aluminum layer acts like a mirror) and then printed so that the laser beam does not pass through it, but is reflected instead.

CD

Thermosets are hard and tightly cross-linked in all spatial directions. They are not plastically deformable, cannot be melted and are highly heat-resistant. Because thermosets are very densely cross-linked, they cannot be dissolved and are very difficult to swell. At room temperature, they are hard and brittle. Plug sockets, for example, are made of thermosets.

thermosets

elastomers

Elastomers (Greek: elastiko = springy; meros = part) are non-meltable, insoluble, but swellable. They have a wide-meshed spatial cross-linking and are therefore in an elastically soft state at room temperature. Examples of elastomer applications are sealing rings or tires.

■ 1.4 How are Plastics Identified?

DIN EN ISO 1043-1

According to the international standard DIN EN ISO 1043-1, plastics are designated by character sequences (abbreviations) that indicate their chemical structure. Additional letters (codes) indicate the application, fillers, and basic properties such as density or viscosity according to DIN EN ISO 1043-2 and DIN EN ISO 1043-3. An example is given in Table 1.1.

HDPE

Table 1.1 Example of Standardized Plastic Identification

Identification of the plastic:
HDPE
Material name:
Linear high-density polyethylene
Abbreviation of the basic polymer product:
PE = polyethylene
Code letters of the additional properties:
H = first code letter for special properties: H = high
D = second code letter for special properties: D = density

example: PC

The CD is made of polycarbonate (PC). PC is a thermoplastic classified according to DIN EN ISO 7391.

In the designation "PC, MLR, 61-09-3", PC stands for polycarbonate, M stands for the injection molding process, L refers to the light and weather stabilizer, and R stands for a mold release agent. The number sequence 61 stands for the viscosity (59 ml/g), the number sequence "0" stands for the melt volume flow rate (MVR 300/1.2 of 9.5 cm^3/g) and the number 3 denotes the impact strength (35 kJ/m^2).

quantities and values

The various quantities and values given here are only to be noted for the time being. Perhaps after reading this book, you will read this section again to see if you can correctly classify many of the previously unknown terms such as "molding compounds" or "MFR value" (melt flow rate), which describes the flowability of the plastic.

1.5 What are the Physical Properties of Plastics?

Plastics are Lightweight

Plastics are typically lightweight construction materials, usually lighter than metals or ceramics. Because some plastics are lighter than water, they can float on the surface. They are used as lightweight components in the construction of airplanes, in automobile production, and for packaging or sports equipment. For example, aluminum is three times as heavy, and steel eight times as heavy, as polyethylene (PE). *(lightweight construction materials)*

The CD spins at a speed of 200 to 500 revolutions per minute. In order for the motor of the CD player to start up quickly and to be small, it is important that the CD is lightweight. *(CD)*

Plastics Are Easy to Process

The processing temperature of plastics ranges from room temperature to approximately 250 °C (482 °F) and in some special cases even up to 400 °C (752 °F). Due to this low processing temperature, which for steel is over 1400 °C (2552 °F), processing is not very complex and relatively little energy is required. This is one reason for the rather low production costs, even for complicated parts. The various processing methods such as injection molding or extrusion will be presented in detail later. *(processing temperature)*

The Properties of Plastics Can be Selectively Optimized

The low processing temperature facilitates the incorporation of additives of various kinds, such as colorant, fillers (e.g., wood flour, mineral powder), reinforcing agents (e.g., glass or carbon fibers) and blowing agents for the production of foamed plastics. *(additives)*

Colorants enable the material to be colored. This eliminates the need for subsequent painting or varnishing in most cases. *(colorants)*

Inorganic fillers in the form of powder and sand can be used in a high proportion (up to 50%). They increase the modulus of elasticity and compressive strength and help to make the plastic more cost effective in many cases. Organic fillers such as (textile) woven fabrics or cellulose webs increase the toughness. Carbon black is incorporated, for example, into car tires (elastomers!). It improves the mechanical properties (abrasion resistance), increases thermal conductivity and light resistance. Incorporation of plasticizers (certain esters and waxes) can change the mechanical behavior of hard plastics to an elastomer-like state. *(fillers)*

reinforcements	Glass, carbon, and aramid fibers, for example, are used as reinforcing materials. They are applied in various forms, e. g., as short or long fibers, as woven fabrics, or mats. The incorporation of specific fibers can boost strength and stiffness several times over.
blowing agents	The use of blowing agents produces synthetic foams whose density can be reduced to 1/100 of the starting material. Foams have particularly effective insulating properties and allow the production of ultra-lightweight components.

Plastics Possess Low Conductivity

insulation	Plastics not only insulate electrical current, as in the case of electricity cables, but also insulate against cold or heat. Examples are a refrigerator or a plastic cup. Plastic's thermal conductivity is about 1,000 times lower than the thermal conductivity of metals.
electrical conductivity	The reason for the poorer conduction of plastics in comparison to metals is that they have practically no free electrons. In metals, these electrons are responsible for transporting heat and electricity. It is precisely this property of plastics that can be significantly influenced by additives.
thermal conductivity	Plastics therefore make suitable insulation materials. However, their low thermal conductivity leads to problems during processing because, for example, the melting heat is transported only very slowly into the interior of the material.

Because of their high insulating properties, plastics can become electrostatically charged. If conductive substances, such as metal powder, are added to the plastic before processing, the insulating effect decreases and with it the tendency to electrostatic charging.

Plastics are Resistant to Many Chemicals

corrosion	The bonding mechanism of atoms in plastics is very different from that of metals. For this reason, plastics are not as susceptible to corrosion as metals. Some plastics are very resistant to acids, bases, or aqueous salt solutions. However, many are soluble in organic solvents such as gasoline or alcohol.
CD CD-ROM DVD solvents	Optical storage media, like CDs, CD-ROMs or DVDs, should therefore not be cleaned with turpentine if they become dirty, as it could damage the plastic. When dissolving plastics, the best solvents are those that have a similar chemical composition to the plastics. As the saying goes, "like dissolves like".

Plastics are Permeable

diffusion material values	The penetration of a substance, e. g., a gas, through another material is called diffusion. The high permeability to gases, resulting from large distances between molecules or a low density, can be disadvantageous. However, permeability differs from

one plastic to another. Permeability does have certain practical applications, such as membranes for seawater desalination plants, for certain packaging films or, for example, organ replacement. Suitable plastics for a particular field of application can be found via such material values as the density, e.g., listed in the manufacturer's specifications or data sheets.

Plastics are Recyclable

Plastics can be reused or processed into other products using various methods. This is referred to as recycling. If recycling does not prove to be economical, various plastics can also be incinerated to generate energy. *recycling*

However, incineration of some substances is problematic and requires specific incineration technology as well as special filter technology. Particularly in the case of plastics that contain chlorine (such as PVC) or fluorine (such as PTFE, better known under the trade name Teflon, for example), the gases produced must be extracted and filtered accordingly. In the meanwhile, labeling of plastic products is obligatory. This makes it possible to identify which plastic the product was made of when being recycled. It is thus possible to sort the waste according to type and recycle it in a specific way. *incineration product labeling*

Additional Characteristics of Plastics

Some plastics are flexible. Although the modulus of elasticity and strength of plastics are wide-ranging, they are usually much lower than the corresponding properties of metals. In many cases, the high degree of flexibility is an advantage for production and application. *flexibility*

Several plastics have better impact strength, compared to mineral glass, with equally good optical properties. This means that plastics do not break as quickly as glass, but in return they are not as scratch-resistant. For this reason, plastics are increasingly replacing glass, for example in civil engineering and in the automotive industry or in the field of optics. *impact strength*

In the case of transparent plastics, in addition to better impact strength, the lower weight also offers an advantage over mineral glass. In automotive engineering, this not only saves weight but allows the vehicle's center of gravity to be lowered. Plastic lenses are more comfortable to wear than glass lenses. *transparency*

1.6 Performance Review – Lesson 1

No.	Question	Answer Choices
1.1	Plastics are divided into the groups consisting of thermoplastics, elastomers and _____.	monomers thermosets
1.2	Thermoplastics are divided into two subgroups: amorphous thermoplastics and _____ thermoplastics.	thermosetting semicrystalline
1.3	Thermoplastics are _____.	meltable non-meltable
1.4	Thermosets are strongly cross-linked and therefore they are non-meltable and _____.	soluble insoluble
1.5	Elastomers are _____ cross-linked.	densely loosely
1.6	Elastomers are _____.	meltable non-meltable
1.7	Most plastics are _____ than metals.	lighter heavier
1.8	The processing temperature of plastics is _____ than that of metals.	higher lower
1.9	Different plastics show _____ degrees of permeability to gases.	identical different
1.10	Plastics are very _____ insulators for heat and electrical current.	poor good
1.11	Most plastics _____ be recycled.	can be cannot be
1.12	The compact disc (CD) is made from the transparent plastic _____.	polyethylene (PE) polycarbonate (PC)

2 Lesson
Raw Materials and Polymer Synthesis

Subject Area	Plastics Chemistry
Key Questions	What raw materials are plastics made of?
	What are the steps of refining from crude oil to the basic substances of plastics?
	How are plastics structured?
	What is a monomer?
	What are macromolecules and what are chain units?
	What methods of polymer synthesis exist?
Contents	2.1 Raw Materials for Plastics
	2.2 Monomers and Polymers
	2.3 Polyethylene Synthesis
	2.4 Methods of Polymer Synthesis
	2.5 Performance Review – Lesson 2
Prerequisite Knowledge	Plastics Fundamentals (Lesson 1)

■ 2.1 Raw Materials for Plastics

The raw materials for plastics production are natural substances such as cellulose, coal, crude oil, and natural gas. The molecules of all these raw materials contain carbon (C) and hydrogen (H). Oxygen (O), nitrogen (N) or sulfur (S) might also be involved. The most important raw material for plastics is crude oil.

carbon chemistry

Figure 2.1 shows the proportion of the various products made from crude oil as a percentage of total crude oil production. It is evident that only six percent of total petroleum is processed into plastics.

crude oil

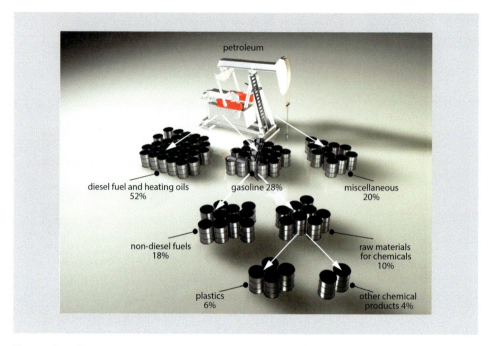

Figure 2.1 Breakdown of raw material products (source: Läpple 2011)

intermediate steps

However, plastics are not produced directly from crude oil. Several intermediate steps are required.

distillation

In a refinery, crude oil is separated into its various components by distillation (a process for separating liquids). For the separation, the differences in boiling points of the various components are exploited. The following are separated: gas, gasoline, petroleum, gasoil, and the residue on distillation is bitumen, which is used in road construction.

cracking

The most important distillate for plastics production is crude gasoline. In a further thermal separation process, the distilled gasoline is broken down into ethylene, propylene, butylene, and other hydrocarbons. This process is also known as cracking. The proportions of the individual cracked products can be controlled via the process temperature. At 850 °C (1562 °F), for example, more than 30% ethylene is obtained.

basic substances

Styrene and vinyl chloride, for example, can also be obtained from ethylene in downstream reaction steps. Like ethylene, propylene, and butylene, these two substances are basic substances (monomers) from which plastics can be produced.

It is well known that all work processes require a certain amount of energy (pressure, heat, motor power, etc). Figure 2.2 shows how energy-efficiently plastic products are manufactured compared to other materials. The graph shows the energy input (calculated in tons of crude oil) required to manufacture items such as pipes and beverage bottles.

energy expenditure

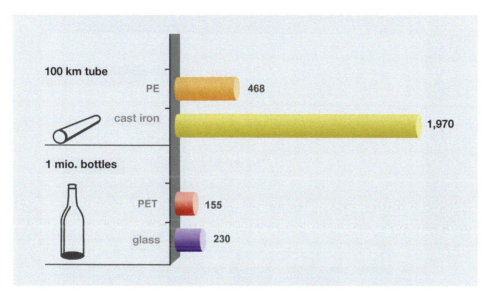

Figure 2.2 Energy required to produce pipes and bottles compared to classic materials (cast iron and glass)

■ 2.2 Monomers and Polymers

The basic substances of plastics are called monomers (Greek: mono = single; meros = part). The plastic macromolecules can be manufactured from these basic substances. The term macromolecule is derived from the size of the plastic molecules (Greek: makros = large), since they result from the combination of many thousands of monomer molecules.

monomers
macromolecules

Prior to the formation of the macromolecule, the monomers exist individually (Figure 2.3). The synthetic material made up of many of these particles is called a polymer (Greek: poly = many). Only by means of a chemical reaction will the individual molecules become a macromolecule.

polymer

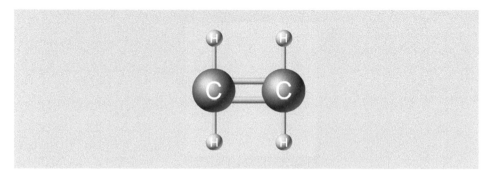

Figure 2.3 Monomer molecule (schematic example of ethylene)

chain elements

Since macromolecules are created from many identical monomers in the simplest case, they consist of identical chain elements that are continuously repeated (Figure 2.4).

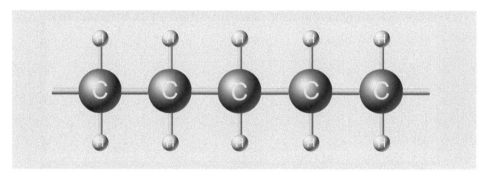

Figure 2.4 Macro molecule (chain units – example of polyethylene)

backbone

Each molecular chain has a continuous line of chain units to which others are attached that are not located in this line. This continuous line of the macromolecule, called the backbone, is most often composed of nothing but the element carbon (C). Oxygen (O) or nitrogen (N) sometimes may also be present. Carbon has the characteristic that it easily forms chains with itself and with oxygen and nitrogen. With other chemical elements, this property is not as prominent.

side chains

Attached to the backbone are various other elements or groups of elements, for example hydrogen (H). If the groups of elements consist of chain-building blocks that actually form a molecular chain, they are referred to as branches or side chains. These branches occur to a greater or lesser extent in all plastics.

2.3 Polyethylene Synthesis

One example of a macromolecular substance is polyethylene (Figure 2.5).

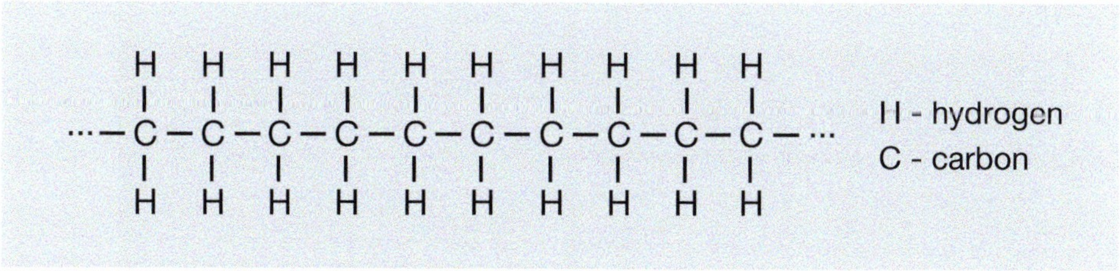

polyethylene

$$\cdots -\underset{\underset{H}{|}}{\overset{\overset{H}{|}}{C}}-\underset{\underset{H}{|}}{\overset{\overset{H}{|}}{C}}-\underset{\underset{H}{|}}{\overset{\overset{H}{|}}{C}}-\underset{\underset{H}{|}}{\overset{\overset{H}{|}}{C}}-\underset{\underset{H}{|}}{\overset{\overset{H}{|}}{C}}-\underset{\underset{H}{|}}{\overset{\overset{H}{|}}{C}}-\underset{\underset{H}{|}}{\overset{\overset{H}{|}}{C}}-\underset{\underset{H}{|}}{\overset{\overset{H}{|}}{C}}-\underset{\underset{H}{|}}{\overset{\overset{H}{|}}{C}}-\underset{\underset{H}{|}}{\overset{\overset{H}{|}}{C}}- \cdots$$

H - hydrogen
C - carbon

Figure 2.5 Structure of a polyethylene linear molecular chain

The monomer from which polyethylene is derived is called ethylene. It consists only of carbon and hydrogen, as shown in the structural formula in Figure 2.6.

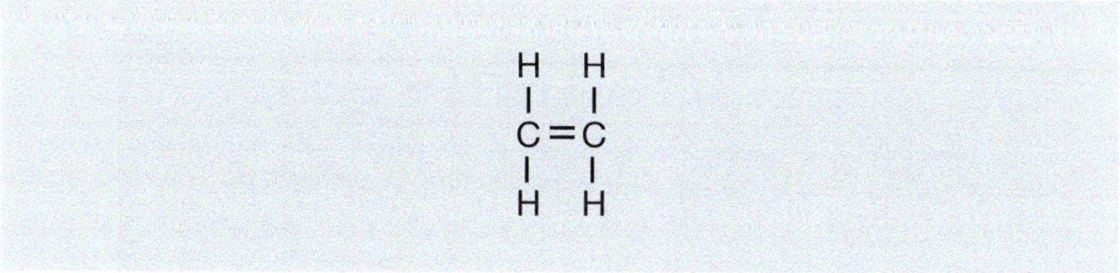

$$\underset{\underset{H}{|}}{\overset{\overset{H}{|}}{C}}=\underset{\underset{H}{|}}{\overset{\overset{H}{|}}{C}}$$

Figure 2.6 Structural formula of ethylene (monomer of polyethylene)

The lines in the figure represent the bonds between the atoms. One bond consists of a pair of electrons. The double lines between the carbon atoms represent a double bond.

bond

The double bond is important for the reaction to form a macromolecule. The ethylene molecules are activated one after another and gradually form a macromolecule, whose structural formula is shown in Figure 2.7.

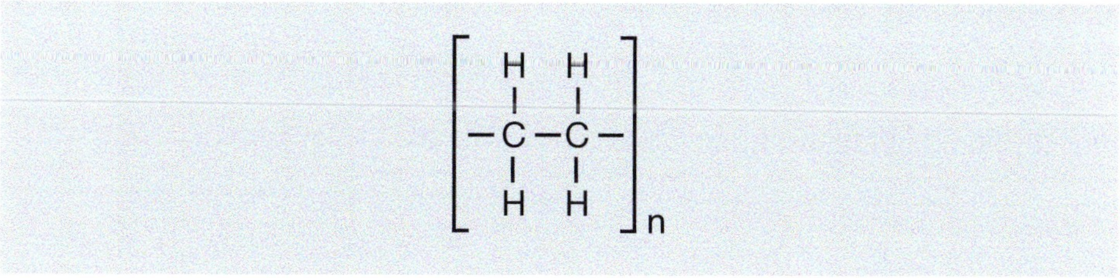

double bond

$$\left[-\underset{\underset{H}{|}}{\overset{\overset{H}{|}}{C}}-\underset{\underset{H}{|}}{\overset{\overset{H}{|}}{C}}- \right]_n$$

Figure 2.7 Structural formula of polyethylene (PE)

tangling

The letter "n" stands for the number that indicates how often this unit is repeated in a macromolecule. It is usually larger than 10,000. Since many macromolecules are formed at the same time rather than one after the other, they become entangled (Figure 2.8). The resulting material is a plastic (Figure 2.8).

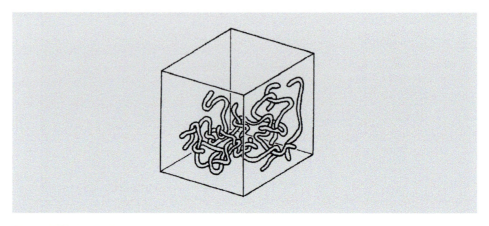

Figure 2.8 Tangled macromolecules

process

Depending on the type of monomer, the macromolecules are formed by different reactions. Plastics are manufactured by means of three fundamentally different reactions or processes. These processes are called synthesis processes, because a new material (here: a plastic) is synthesized (Greek: synthesis = combine) from building blocks (here: monomers). The technical term for the plastics produced by these synthesis processes is based on the name of the particular process (Table 2.1).

Table 2.1 Product Names and Examples of Plastics

Synthesis process	Product name	Example	Abbreviation
polymerization	addition polymers	polyethylene	PE
		polypropylene	PP
polyaddition	polyaddition products	epoxy resins	EP
		polyurethane	PUR
polycondensation	condensation polymers	polyamide	PA
		polycarbonate	PC

PE, PP

Polyethylene (PE) and polypropylene (PP) are so-called commodity plastics, which are extensively used because of their low price and wide range of applications. Products made of PE and PP can also be manufactured using a variety of plastics processing methods. Examples include protective and packaging films, bottles, pipelines, transport containers, electrical accessories, covers, fittings and housings;

they are also used in washing machines, in electrical installations and in the manufacturing of fittings, containers and apparatus.

Epoxy resins (EP) serve as matrix material in the field of fiber-reinforced polymers (FRP), also called fiber composites. They are also used as coatings, e.g., in civil engineering or tank construction, as they are highly resistant to acids, for example. EP

Polyurethane (PUR) is used for shoe soles, rollers, and bearings, as well as in foamed form as an insulating material or cushioning foam plastic, for example in mattresses. PUR

Polycarbonate (PC) is a slightly expensive material, but because of its very good transparent properties it is often used as a glass substitute. Examples of such applications include automotive parts (windows, headlight covers) and eyeglass lenses. Another excellent example is the CD, manufactured from polycarbonate. PC

■ 2.4 Methods of Polymer Synthesis

Three polymer synthesis processes are distinguished for the production of plastics: addition polymerization, polycondensation and polyaddition. These processes are presented and outlined in the following subsection.

Addition Polymerization

The double bond, which is present between two carbon atoms in the monomer, plays a decisive role in polymerization. We will show this with the example of vinyl chloride (Figure 2.9). double bond

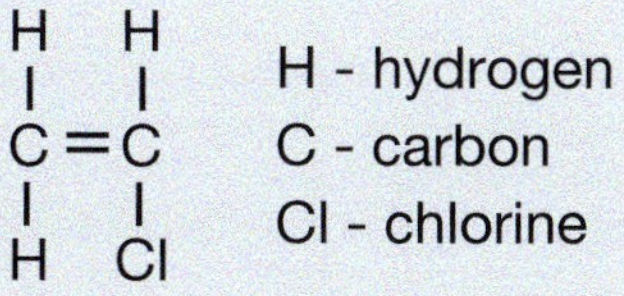

$$\begin{array}{cc} H & H \\ | & | \\ C = C \\ | & | \\ H & Cl \end{array} \quad \begin{array}{l} H - \text{hydrogen} \\ C - \text{carbon} \\ Cl - \text{chlorine} \end{array}$$

Figure 2.9 Structural formula of vinyl chloride

In the case of vinyl chloride, as well, each bond of the molecule consists of two electrons. The double bond thus consists of two bonds with two electrons each. One of the two bonds in the double bond can be cleaved rather easily. In other words, it can be cleaved into two single (unpaired) electrons. double bond cleavage

macromolecules
radical

Bond cleavage leads to the formation of the molecular chain. It starts with the cleavage of a double bond, which is caused by another particle, e.g., a radical. Radicals are highly reactive elements or groups of elements. This means that they react very readily with other molecules. The reason for this is a free single electron that each radical possesses and that readily forms a bond with another electron. The cleavage of the double bond in vinyl chloride by a radical (R) is shown in Figure 2.10.

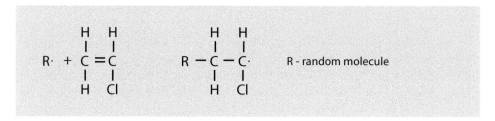

Figure 2.10 Cleavage of the double bond

chain formation

During bond cleavage, the electron of the radical forms a new bond with one electron of the cleaved bond. The other electron of the cleaved bond is now located on the other side of the vinyl chloride. This side with its free electron can now cleave new double bonds again. Thus, a long chain is formed from that beginning (Figure 2.11).

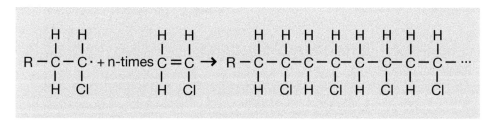

Figure 2.11 Chain formation

length of chains

This growth ends when two chain ends or a chain end and a radical meet. But since there are initially many more vinyl chloride monomers than chain ends or radicals, the chains become very long before they stop growing. To determine the properties of the plastic, the length of these chains is of great importance. The length is specified as the number "n" of repeating chain units (Figure 2.12).

polymers

The number "n" is usually more than 10,000. To get an idea of how long such a macromolecule can be, try to imagine the molecule enlarged 1,000,000 times. It would then be 20 cm (8 in) thick and already 1 km (3300 ft) long. Plastics that are produced by addition polymerization are called polymers (Table 2.2).

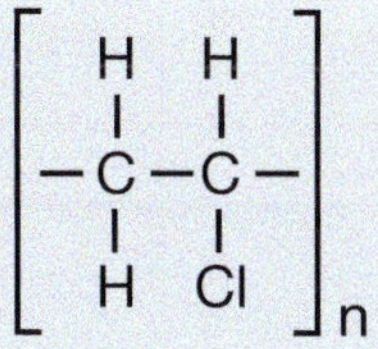

$$\left[\begin{array}{c c} H & H \\ | & | \\ -C - C - \\ | & | \\ H & Cl \end{array} \right]_n$$

Figure 2.12 Repeat unit

Table 2.2 Addition Polymerization and Applications

Polymers	Abbreviation	Products
polyethylene	PE	protective and packaging films, bottles, pipes, transport containers, electrical accessories, covers, fittings, chemical apparatus engineering
polypropylene	PP	instrument housings, washing machine parts, electrical installations, pipes, fittings, apparatus engineering
polymethyl methacrylate	PMMA	glazing, rear lights, sanitary parts, signs, lenses, drawing equipment, light domes

How can you memorize the polymerization process? A train can only be coupled together if there is a coupling at the front and rear of each wagon. Similarly, a macromolecule chain is formed during polymerization by coupling the individual monomers together by means of the electrons of the cleaved double bond. The catchword for polymerization is therefore "coupling". — catchword "coupling"

One or more types of monomers can be used simultaneously to manufacture a plastic by addition polymerization. If only one monomer is used in polymerization, the result is a homopolymer. If the polymer made from two or more different monomers, this is called copolymerization (co = with, together) and a copolymer is obtained. The configuration of the different monomer building blocks in the copolymer can be diverse. The properties of the plastic can be influenced via the choice of the different monomers in copolymerization. — copolymer

Polycondensation

A typical feature of the polycondensation reaction is that small molecules, mostly water molecules, are cleaved off. This process of cleavage off is called condensation in organic chemistry. Hence the name of this type of plastic synthesis. The chemical formula for water is H_2O. A water molecule is thus composed of two hydrogen atoms (H) and one oxygen atom (O). — polycondensation

The formation of macromolecules by means of the polycondensation reaction requires molecules that possess two or more so-called "functional" groups (Figure 2.13).

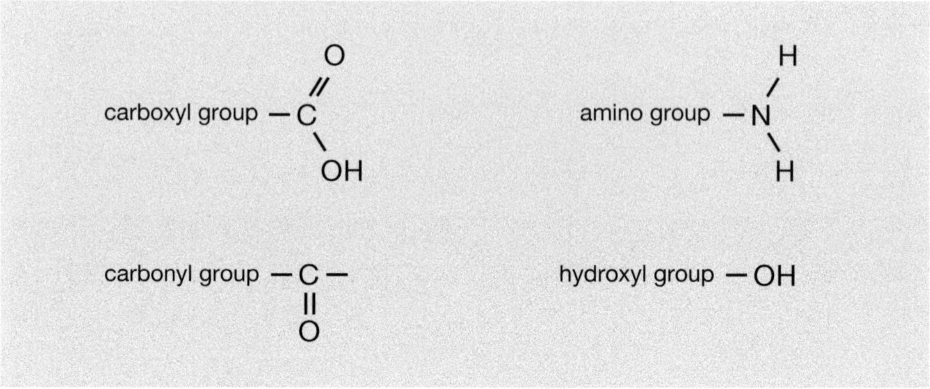

$$
\begin{array}{ll}
\text{carboxyl group} - C\!\!\!\begin{array}{c}\scriptstyle O\\ \scriptstyle \|\\ \end{array}\!\!\!-OH & \text{amino group} -N\!\!\!\begin{array}{c}H\\ H\end{array} \\[2ex]
\text{carbonyl group} -\underset{O}{\overset{\|}{C}}- & \text{hydroxyl group} -OH
\end{array}
$$

Figure 2.13 Functional groups

However, a bond can only be formed between two molecules if two different functional groups are present, from which the particles cleaved off and then "condense" as water.

types of molecules

Thus, to ensure the reaction can form a continuous chain, we must have the following types of molecules in polycondensation: either one type of molecule that has at least two different functional groups, or at least two different molecule types that each have two or more of the same functional groups.

polyamide

One example of polycondensation is the reaction in which an amide is obtained from two molecules. The resulting plastic made from numerous molecules is therefore called polyamide. An example of polycondensation is the reaction between hexamethylene diamine and adipic acid (Figure 2.14) which results in polyamide 66.

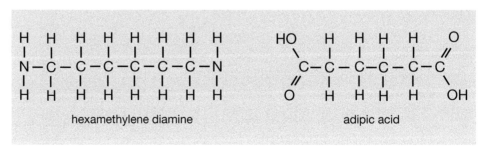

Figure 2.14 Structural formulas

The reaction proceeds in two steps, with the particles cleavage off from the functional groups in the first step. In the second step, the macromolecule polyamide and water are obtained (Figure 2.15). Since a reverse reaction is also possible, the water that is cleaved off must be constantly removed during the production of polyamide.

reaction process

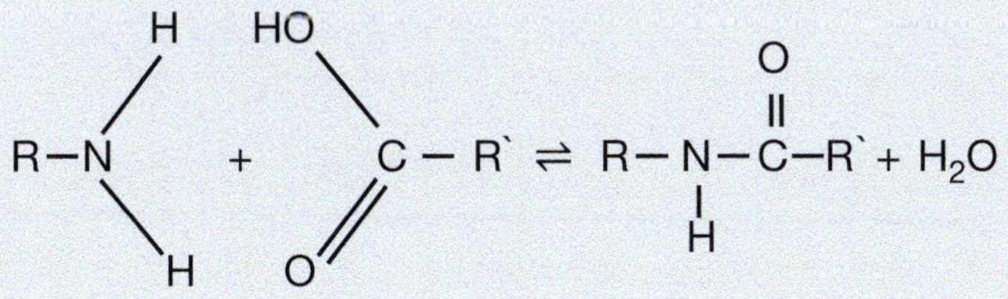

$$R-N\overset{H}{\underset{H}{\diagup}} + \overset{HO}{\underset{O}{\diagup}}C-R' \rightleftharpoons R-\underset{H}{\overset{}{N}}-\overset{O}{\overset{\|}{C}}-R' + H_2O$$

Figure 2.15 Reaction sequence for the formation of polyamide

The molecules cleaved off during the reaction, in this case the water, must be constantly removed so that the reaction can continue and form very long chains. As with polymerization, there is no definitive termination step. Plastics resulting from polycondensation are called condensation polymers (Table 2.3).

water removal
condensation polymers

Table 2.3 Condensation Polymerization and Applications

Condensation polymers	Abbreviation	Products/examples
phenol-formaldehyde (resin)	PF	handles for gearshifts, switch parts, car ashtrays, heaters, irons, pots and pans, lamp sockets
unsaturated polyester	UP	glass fiber reinforced used in boat manufacturing, vehicle construction, equipment housings, rotor blades for wind turbines
polycarbonate	PC	housings for office and household machinery, observation glasses, CDs, DVDs, Blu-Ray discs, camera housings, signal lighting
polyamides	PA	gears, slide rollers, housings for electrical devices, plugs

Our CD is also made of a plastic produced by polycondensation, polycarbonate (PC).

CD

How to memorize the polycondensation process? In the polycondensation process, water is separated. So, the catchword for polycondensation is "separation".

catchword "separation"

polyaddition

Polyaddition (Step Addition)

A polyaddition reaction proceeds similarly to that of polycondensation. The difference is that here no particles are cleaved off, but a hydrogen atom migrates from one functional group to another.

Thus, as in polycondensation, two different functional groups are required to form a bond. The monomers used must again each have at least two functional groups. Again, either one type of molecule with at least two different functional groups or at least two types of molecules, each with two or more identical groups, are used to form the macromolecules.

reaction sequence

The reaction can be represented in three steps:

Step 1: There is one molecular end with an easily cleavable hydrogen atom and one molecular end with an easily cleavable bond.

Step 2: The hydrogen atom is separated, and the bond of the other functional group is cleaved.

Step 3: The hydrogen atom forms a bond with one of the electrons of the cleaved bond. The site from which the hydrogen atom has cleaved off and the other electron of the cleaved bond form a new bond; the chain is thereby lengthened.

Figure 2.16 shows a schematic representation of the polyaddition reaction.

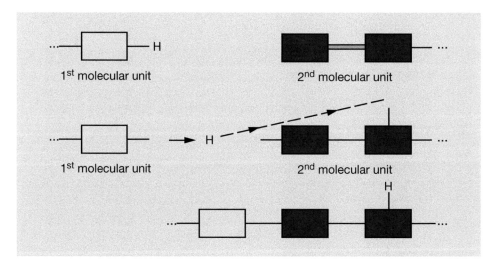

Figure 2.16 Polyaddition reaction

polyaddition products

The polymers obtained from the polyaddition reaction are called polyadducts. In Table 2.4, some polyaddition products are presented in their applications.

Table 2.4 Polyaddition Products and Applications

Polyaddition products	Abbreviation	Products/examples
polyurethane	PUR	shoe soles, rollers, bearings, clutch disks
polyurethane foams	PUR foam	insulating and cushioning foams for furniture, buildings, clothing
epoxies	EP	adhesives, coatings for containers, also fiber reinforced for tools

The characteristic feature of this chemical reaction is the "exchange" of an atom, which changes from the functional group of one reaction partner to the functional group of the other reaction partner.

catchword "exchange"

■ 2.5 Performance Review – Lesson 2

No.	Question	Answer Choices
2.1	The raw materials coal, crude oil and _____ are used to produce plastics.	steel aluminum natural gas
2.2	Starting materials for plastics are obtained from crude oil through the processing steps of distillation and _____.	conversion cracking
2.3	Monomers for plastics produced from petroleum include ethylene, vinyl chloride and _____.	kerosene propylene
2.4	You can think of a macromolecule as a long _____.	chain cord
2.5	Monomer is the chemical name for molecules from which the plastic is made. The plastic, which is composed of many such individual parts, is therefore called _____.	thermoset copolymer polymer
2.6	In most cases, the chemical element _____ forms the continuous line (backbone) of macromolecules.	carbon (C) oxygen (O) nitrogen (N)
2.7	The abbreviations of plastics often contain the letter P in the first place. It stands for _____.	partial poly particle
2.8	After the macromolecules have been formed, they exist in a _____ state.	tangled stretched

No.	Question	Answer Choices
2.9	Polyethylene (PE) has a very simple structure. It is only composed of the elements hydrogen (H) and _____.	carbon (C) fluorine (F) nitrogen (O)
2.10	Polycarbonate (PC) is used as a glass substitute, particularly because of its good _____ properties.	electrical acoustical optical
2.11	During polymerization, the _____ plays the decisive role.	triple bond double bond
2.12	The catchword for addition polymerization is _____.	"coupling" "exchange" "separation"
2.13	Plastics produced by addition polymerization are, for example, polyethylene (PE) and _____.	polycarbonate (PC) polypropylene (PP)
2.14	The catchword for polycondensation is _____.	"exchange" "separation" "coupling"
2.15	The catchword for polyaddition is _____.	"coupling" "separation" "exchange"

Lesson 3
Classification of Plastics

Subject Area	Physics of Plastics
Key Questions	Into which groups can plastics be divided?
	What criteria are used to classify them?
	Which primary processing methods are applied to plastics?
	What are the fabrication processes for plastics?
	What are the shaping processes for plastics?
Contents	3.1 Bonding Forces in Polymers and Their Temperature Behavior
	3.2 Identification of Categories of Plastics
	3.3 Thermoplastics
	3.4 Cross-Linked Plastics (Elastomers and Thermosets)
	3.5 Fabrication and Processing Methods
	3.6 Methods for Shaping Thermoplastics
	3.7 Performance Review – Lesson 3
Prerequisite Knowledge	Plastics Fundamentals (Lesson 1)
	Raw Materials and Polymer Synthesis (Lesson 2)

3.1 Bonding Forces in Polymers and Their Temperature Behavior

Bonding Forces within Molecules

atomic bonds

The atoms of the monomer molecules that compose the macromolecules are linked by atomic bonds, also called covalent bonds. These bonds can be understood as forces that hold two atoms together. Generally, in illustrations showing molecules, the bonds are represented with lines. One example is the monomer ethylene (Figure 3.1).

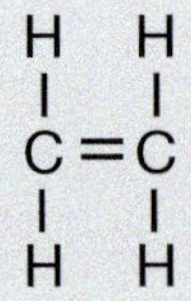

$$\begin{array}{c} H \quad H \\ | \quad | \\ C = C \\ | \quad | \\ H \quad H \end{array}$$

Figure 3.1 Structural formula of ethylene (H = hydrogen atom, C = carbon atom)

number of bonds

Depending on the number of bonds between two atoms, a distinction is made between single, double and triple bonds. As shown above, ethylene contains a double bond between the two carbon atoms (C) and a single bond between each hydrogen atom (H) and a carbon atom. The double bond is an unsaturated bond. Unsaturated means that the bond can be easily cleaved. This makes it possible to form another bond with other atoms. These bonding forces also occur in macromolecules of the plastic.

Intermolecular Bonding Forces

intermolecular forces

Forces exist not only within a molecule, but also between adjacent molecules. These latter forces are therefore called intermolecular forces. They cause two molecules to attract one another with a certain force, such that they cannot move away from one another on their own (Figure 3.2).

strength

These intermolecular forces between the tangled macromolecules also operate within the plastic. They give the plastic most of its strength since they hold the molecules together and prevent them from easily "slipping" away from each other. The bonds can be thought of as the hooks in a Velcro fastener. The hooks cause the strips of fabric to hold each other tightly. They can only be separated if they are pulled hard.

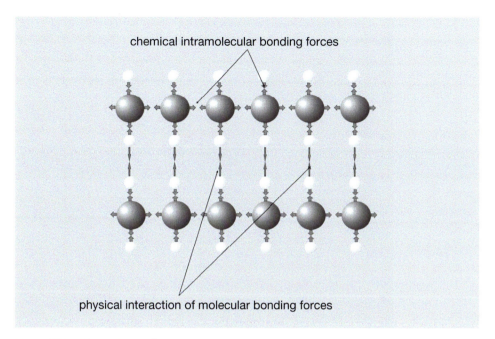

Figure 3.2 Intermolecular forces

However, intermolecular bonds are not as strong as atomic bonds. When subjected to a load, the bonds between the molecules are the first to be broken. Let's look at this again using the example of Velcro. Imagine that the atomic bonds are the forces that hold the woven strip of fabric together. So, if we pull hard, the fabric strip doesn't tear first, but instead the hooks separate, and the fabric slips off. The intermolecular bonds are the first to separate.

atomic bonds

Effects of Temperature

Heat is manifested in the movement of molecules. The higher the temperature, the more the molecules move. This movement reduces the intermolecular forces. Above a certain temperature, these forces are completely eliminated, allowing free movement of the molecules that previously had been connected by these forces. If the temperature drops again, the movement of the molecules is reduced in turn and the forces develop again.

heat

The covalent bonds between the atoms of a molecule are not broken by the movement that happens under heat. They are much stronger and are only destroyed at much higher temperatures. Unlike the intermolecular forces, they do not redevelop as the temperature drops. The molecule remains destroyed.

thermal motion

Another consequence of the increasing movement of the molecules is that they require more space due to this movement. The plastic therefore expands with in-

thermal expansion

creasing temperature. However, this change in volume with the change in temperature, known as thermal expansion, varies from one material to another. Different plastics have different degrees of thermal expansion. A measure of the change in length is the coefficient of linear thermal expansion. The higher it is, the more the material expands under heat (Table 3.1).

Table 3.1 Coefficient of Thermal Expansion of Various Materials

Material	Abbreviation	Coefficient of linear thermal expansion (1/K 10^{-6}) at 50 °C (122 °F)
polyethylene	PE	150 – 200
polycarbonate	PC	60 – 70
steel	St	2 – 17
aluminum	Al	23

CD

A CD is made of the plastic called polycarbonate. With a coefficient of linear thermal expansion between 60 and 70, PC has about 3 times less expansion than polyethylene (PE), which is used to make packaging films, for example.

■ 3.2 Identification of Categories of Plastics

plastic categories

As we have shown, there are different bonding forces in plastics. Plastics are classified according to macromolecular structure and the nature of the bonding mechanisms.

obsolete terms

The four categories composed of amorphous thermoplastics, semicrystalline thermoplastics, thermosets, and elastomers are described in more detail below. The literature often contains other, obsolete terms. For example, thermosets are sometimes also referred to as "duromers" and thermoplastics as "plastomers".

■ 3.3 Thermoplastics

definition

Plastics consisting of macromolecules with linear or branched chains that are held together by intermolecular forces are called thermoplastics. The strength of these intermolecular forces depends, among other factors, on the type and number of branches or side chains (see Figure 3.3).

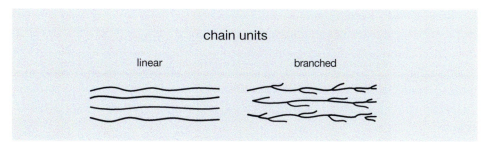

Figure 3.3 Linear and branched chain molecules

The term thermoplastics is derived from the words thermos (= hot, heat) and plastikos (= shapeable, moldable) since the intermolecular forces in thermoplastics weaken under heat and thus make them moldable.

thermos

Amorphous Thermoplastics

Plastics composed of highly branched molecular chains and long side chains, on the other hand, cannot assume a state of tight packing, even in parts. This is due to their irregular structure. Such chain molecules are intertwined in and around each other like a ball or a wad of cotton. The plastic is unstructured (amorphous). It is therefore referred to as an amorphous thermoplastic.

amorphous

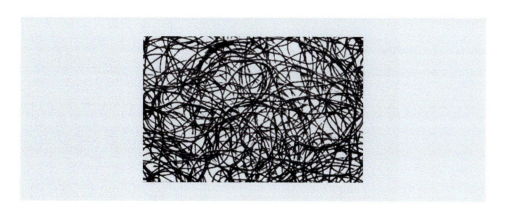

Figure 3.4 Structure of an amorphous thermoplastic

Since amorphous thermoplastics are as transparent as glass in the uncolored state, they are also referred to as synthetic or organic glasses.

as transparent as glass

The "optical data carriers" such as the CD, CD-ROM and the DVD are also made from an amorphous thermoplastic. Because the latter is crystal clear, the laser can scan the depressions (pits) in the plastic, in relation to the reflective aluminum or gold layer, and pass this information to the CD player, which converts it back into music.

CD

Semicrystalline Thermoplastics

If the macromolecules exhibit only slight branching, i.e., a few short side chains, then areas of the individual molecular chains are ordered and thus densely packed together. Such regions with a high state of order of the molecules are called crystalline regions. However, complete crystallization does not occur due to the long molecular chains, which also entwine around and into each other during polymerization.

semicrystalline

Only a certain portion of the molecules assemble in an ordered manner, while other portions are further apart and disordered. These disordered regions are called amorphous regions. Thermoplastics in which both crystalline and amorphous regions exist together are therefore referred to as semicrystalline thermoplastics (Figure 3.5).

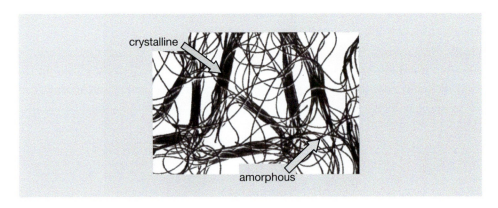

Figure 3.5 Structure of a semicrystalline thermoplastic

cloudy, milky

The semicrystalline thermoplastics are never as transparent as glass, even in the uncolored state. Instead, they are always somewhat cloudy or milky, due to light scattering at the boundaries between the amorphous and crystalline regions of the plastic.

■ 3.4 Cross-Linked Plastics (Elastomers and Thermosets)

cross-links

Besides the group of thermoplastics, there are groups of plastics in which the individual molecular chains are linked together by transverse bonds (bridges). These transverse bonds (bridges) are also referred to as cross-links and, accordingly, the materials are called cross-linked plastics. The groups are distinguished by the number of cross-links and are subsequently subdivided into elastomers and thermosets. The molecules of these materials are thus held together not only by intermolecular forces, but also by atomic bonds.

Elastomers

The chain molecules in elastomers are randomly arranged and exhibit relatively few cross-links. The group of elastomers therefore has a wide mesh cross-linkage. — *characteristics*

Elastomers behave like rubber at room temperature. The cross-links severely limit the mobility of the individual molecular chains relative to one another. As with the atomic bonds in the macromolecules, the atomic bonds in the bridges can only be broken by very high temperatures, and they are not re-established even when the temperature drops. — *properties*

Elastomers therefore do not melt and are not soluble. Elastomers can, however, swell to a certain extent. This is because the molecular chains have only a few cross-links, and therefore other small molecules, such as oil or gasoline, can penetrate between the chains of the elastomer.

Thermoplastic Elastomers (TPE)

Thermoplastic elastomers, which combine the easy processing of thermoplastics with the properties of elastomers, are increasingly penetrating new areas of application. Furthermore, they are replacing traditional materials from the thermoplastic and elastomer sectors.

TPEs can be melted repeatedly and formed precisely during cooling. At room temperature, they exhibit rubber-elastic behavior. In terms of their structure and behavior, they are therefore intermediate between thermoplastics and elastomers. They are most widely used in automotive engineering and in the sports and leisure industry.

Thermosets

Another category is represented by the thermosets, which also have a random arrangement of molecular chains. But compared to the elastomer structure, they have a much higher number of cross-linking sites between the individual molecular chains. Plastics that are made up of such extensively cross-linked chain molecules are called thermosets. — *characteristics*

At room temperature, these extensively cross-linked molecules are very hard and strong, but brittle (i.e. sensitive to impact). They show considerably less softening on heating than thermoplastics. Due to their extensive cross-linking, they are neither meltable nor swellable, just like the elastomers. — *properties*

Figure 3.6 shows the structural arrangement of the macromolecules in elastomers and thermosets. Elastomers, like thermosets, have a random arrangement of molecular chains, but thermosets have a much higher number of cross-linking sites. Therefore, in contrast to elastomers (elastic), they are not elastic but solid and brittle.

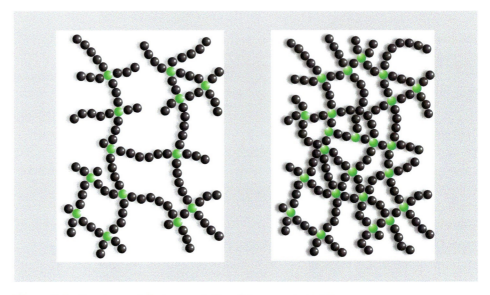

Figure 3.6 Structures of elastomers (left) and thermosets (right)

■ 3.5 Fabrication and Processing Methods

raw plastic

From the plastic produced in chemical processes to the plastic product used by the consumer, several intermediate steps are required. The raw material "plastic" is produced in the form of pellets, as powder, paste or in liquid form and then processed into semi-finished or finished parts.

semi-finished products
finished products

Semi-finished products are intermediate products that are further processed into an end product using various processing techniques, such as forming. Examples of semi-finished products are sheets, films, pipes, and profiles made of plastic. Finished products are end products that are manufactured by primary shaping, such as injection molding. Examples of finished products are beverage crates, gears, and housings.

overview

An overview of the fabrication and processing methods for thermoplastics and thermosets is given in Table 3.2.

Table 3.2 Overview of Fabrication and Processing Methods

Shaping technology	Thermosets	Thermoplastics
primary shaping	molding compounds are shaped with a simultaneous chemical reaction: • thermosetting molding compounds • liquid reactive resin	molding compounds are shaped in a thermoplastic state
forming/thermoforming	–	semi-finished products are formed in the thermoelastic state
machining	shape cutting operations	shape cutting operations
joining	mechanical joining methods as well as bonding	mechanical joining methods as well as bonding and welding

You will notice in this overview that no shaping technology is mentioned for the thermosets – and this also applies to the elastomers. Cross-linked plastics do not exhibit a thermoplastic state range and can therefore no longer be formed after the curing process.

thermosets

Machining is a general term for the shaping of plastics, involving cutting operations such as turning, milling, sawing, etc. The term "cutting" is used to describe these operations of shaping plastics.

machining

Joining processes of plastics, which include bonding and welding as well as the mechanical joining processes of screwing, riveting, etc., are referred to by the generic term joining.

joining

The forming, machining, and joining processes are grouped together under the heading of "fabrication" processes, whereas the primary forming processes, such as extrusion and injection molding, are classified as "processing" methods.

fabrication processing

■ 3.6 Methods for Shaping Thermoplastics

Table 3.3 shows the shaping processes associated with the material state of the thermoplastics.

Table 3.3 Associated Shaping Processes

Shaping process	Material state		
	Rigid	Thermoelastic	Thermoplastic
primary shaping	–	–	extrusion casting calendering injection molding compression molding
forming	–	folding, bending, embossing, knurling, stretch forming, vacuum forming, combined processes	–
machining	drilling, turning, milling, planing, sawing, cutting, grinding	–	–
joining	screwing, riveting, bonding	–	welding

This classification does not exist for cross-linked plastics, thermosets, and elastomers. These plastics can only be used to produce parts that have their final shape after cross-linking or can only be subsequently processed mechanically by joining or separating. These plastics cannot be welded either, as they do not have a thermoplastic state.

■ 3.7 Performance Review – Lesson 3

No.	Question	Answer Choices
3.1	The bonds between atoms within a macromolecule are called _____ or covalent bonds.	intermolecular bonds atomic bonds
3.2	The bonds interacting between two macromolecules are referred to as _____.	intermolecular bonds atomic bonds
3.3	The forces of an atomic bond are considerably _____ than those of an intermolecular bond.	stronger weaker
3.4	Thermoplastics are divided into amorphous and _____ thermoplastics.	thermosetting semicrystalline
3.5	Amorphous thermoplastics are _____ at room temperature.	cloudy crystal clear

3.7 Performance Review – Lesson 3

No.	Question	Answer Choices
3.6	The molecules of thermosets are _____ cross-linked.	extensively lightly
3.7	Elastomers have few cross-linked molecules and are therefore _____ in a solvent.	swellable not swellable
3.8	Optical data carriers (e.g., the CD) are made of an amorphous thermoplastic, because the plastic must be _____.	transparent meltable cross-linked
3.9	Injection molding is a processing method that is classified as _____.	forming primary shaping joining machining
3.10	Thermoforming is a manufacturing process that is classified as _____.	reshaping primary shaping joining machining
3.11	_____ is a method of joining.	Bonding Extrusion Thermoforming
3.12	_____ is a machining process.	Sawing Welding Bonding
3.13	Thermosets and elastomers cannot be _____ because they do not have a thermoplastic stage when heated.	primary shaped reshaped
3.14	Which joining process cannot be used with thermosets? _____.	bonding welding mechanical joining

4 Lesson: Deformation Behavior of Plastics

Subject Area	Physics of Plastics
Key Questions	How do plastics behave when exposed to heat?
	How do amorphous and semicrystalline thermoplastics differ in this regard?
	How do cross-linked plastics, i.e., elastomers and thermosets, behave?
Contents	4.1 Behavior of Thermoplastics
	4.2 Amorphous Thermoplastics
	4.3 Semicrystalline Thermoplastics
	4.4 Behavior of Cross-Linked Plastics
	4.5 Performance Review – Lesson 4
Prerequisite Knowledge	Classification of Plastics (Lesson 3)

■ 4.1 Behavior of Thermoplastics

Deformation behavior describes how the shape of a part changes when subjected to load (force) and temperature. Deformation behavior helps to describe the difference between a semicrystalline and an amorphous thermoplastic.

deformation behavior

Let us now explain tensile strength and elongation at maximum tension in more detail. If a pulling force is applied to a plastic specimen, two things can be observed:

- The plastic specimen withstands a specific, maximum tensile force. The stress at maximum force is called tensile strength. It represents a measure of the strength of the plastic.

tensile strength

elongation at break

- But we also notice during the tensile test that the plastic specimen elongates. It is therefore lengthened. The degree of elongation at which the plastic specimen breaks is referred to as the elongation at break. The toughness of the plastic can be concluded from this measurement.

effects of temperature

Both measured values depend on the temperature at which they are determined. In the following, we will now look at the deformation behavior of different thermoplastic groups.

4.2 Amorphous Thermoplastics

The deformation behavior of an amorphous thermoplastic is shown in Figure 4.1.

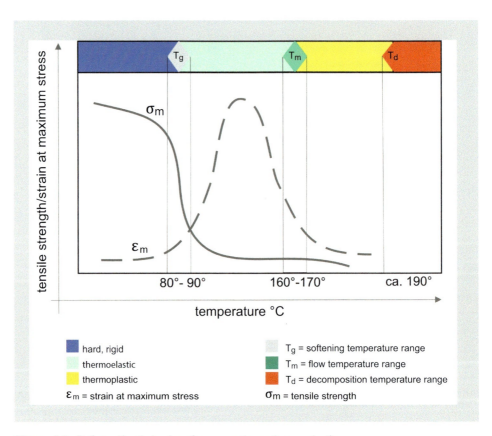

effects of temperature

Figure 4.1 Deformation behavior of an amorphous thermoplastic

At room temperature, the plastic is a rigid material. Since the individual macromolecules hardly move, they are held together by intermolecular forces. If the temperature now rises, the macromolecules move with increasing intensity. The strength of the material decreases and at the same time its extensibility and toughness increase.

After exceeding the softening temperature range (ST), the intermolecular forces have become so weak that external forces can cause the macromolecules to slide away from each other. At T_g the strength declines steeply, while the elongation suddenly increases. In this temperature range, the plastic is in a rubbery or thermoelastic state in which it can be reshaped, for example.
glass transition temperature T_g

With further temperature increase, the intermolecular forces are completely eliminated. The plastic continuously transitions from the thermoelastic to the thermoplastic range. This transition is characterized by the flow temperature range (FT) around the melting temperature T_m. It is not a temperature that can be specified exactly. In the thermoplastic range, plastic pipes, for example, are produced by the extrusion process. The welding of thermoplastics is likewise only possible in the thermoplastic range.
flow temperature range (FT)
extrusion
thermoplastic welding

If the plastic is heated further, its chemical structure will at some point be destroyed. This limit is marked by the decomposition temperature (T_d).
decomposition temperature (T_d)

The optical storage media such as the CD and the DVD, but also the Blu-ray, are manufactured from an amorphous thermoplastic by injection molding (Lesson 10). In contrast to the semicrystalline thermoplastics, the amorphous thermoplastics are transparent and thus the laser can read the information on the optical data carrier (music, speech, image material, etc.). Of the various amorphous thermoplastics, polycarbonate (PC) has been found to be the most suitable material for optical storage media.
CD, DVD, Blu-ray

The upper service temperature limit of the amorphous thermoplastic PC, from which the CD is made, is 135 °C (275 °F). It is therefore possible for a CD to remain fully functional even if it is heated up to 80 °C (176 °F) on the dashboard of a car in direct sunlight.

■ 4.3 Semicrystalline Thermoplastics

As previously discussed, in the case of semicrystalline, as opposed to amorphous, plastics, two regions are present next to each other. One is the crystalline region, in which the molecules are closely packed. The other is the amorphous region, in which the molecules are further apart. The intermolecular forces that hold the crystalline state together are far stronger than those of the amorphous region.
amorphous and crystalline range

The deformation behavior of a semicrystalline thermoplastic is shown in Figure 4.2.

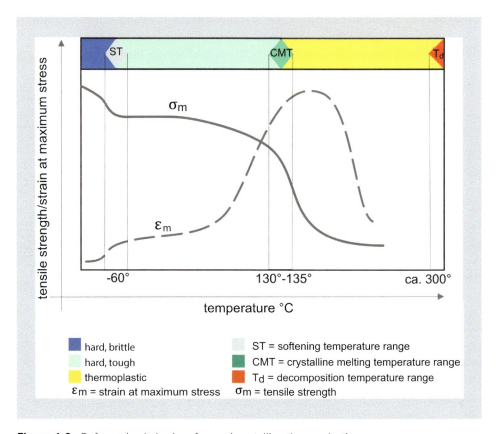

Figure 4.2 Deformation behavior of a semicrystalline thermoplastic

glass transition temperature T_g

All areas of the plastic below the softening temperature range (ST) are hard and brittle. This softening temperature range is also referred to as the "glass transition temperature range" T_g. Within this temperature range, the plastic cannot be used for practical applications.

temperature increase

Once the glass transition temperature range T_g is exceeded, the mobility of the molecular chains first increases in the amorphous regions, since the intermolecular forces here are not as strong as in the crystalline regions, which still remain solid. For common semicrystalline plastics, this temperature is below room temperature. The plastic now exhibits both toughness and strength at the same time.

crystalline melting temperature CMT

As the temperature rises, the mobility of the molecular chains in the amorphous regions continues to increase. In the crystalline regions, the molecules also start to move slowly. Soon the crystalline melting temperature (T_m) is reached. At this point, the intermolecular forces in the amorphous regions of the semicrystalline thermoplastics are completely eliminated. Within the crystalline melting temperature range (CMT), the semicrystalline thermoplastic is thermoelastic and can be reshaped. In contrast to the amorphous thermoplastic, the thermoelastic range is very restricted here and must be maintained very accurately during the reshaping process. Above the crystalline melting temperature range, the bonding forces are too weak to prevent displacement and slippage of the chain molecules even in the crystalline regions of the semicrystalline thermoplastics.

The entire plastic now starts melting within the decomposition temperature range. On further heating, the plastic is destroyed above the decomposition temperature (T_d).

decomposition temperature (T_d)

4.4 Behavior of Cross-Linked Plastics

The deformation behavior of elastomers and thermosets may best be explained with the aid of the torsion pendulum test. In the torsion pendulum test, the shear modulus (G) of the plastic is determined.

shear modulus

The shear modulus is a measure of the rigidity of the plastic. Figure 4.3 shows the shear modulus as a function of temperature for different cross-linked plastics with different degrees of cross-linking.

rigidity

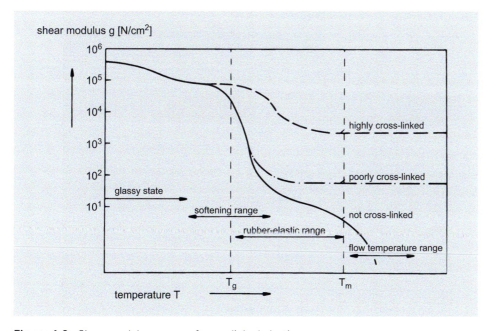

Figure 4.3 Shear modulus curves of cross-linked plastics

softening temperature	In the temperature range below the softening temperature range (ST), the plastic is hard and brittle, regardless of its degree of cross-linking.
elastomer	The shear modulus curve of the poorly cross-linked plastic (elastomer) declines immediately after exceeding the softening temperature range (ST), so that the plastic has only low stiffness.
flow temperature range decomposition temperature	In contrast to thermoplastics, however, the poorly cross-linked plastic retains this stiffness even with a further temperature increase above FT. The reasons for this behavior are the cross-linking sites in the elastomer, which make it impossible for the individual molecular chains to slip away from each other. Thus, the plastic will not melt, but will decompose under a further temperature increase above the decomposition temperature (DT).

Figure 4.4 shows the temperature ranges or physical-state ranges of various plastics relative to those of water. The gaseous phase of water (steam), which can be converted back into water when the temperature is reduced, does not exist for plastics.

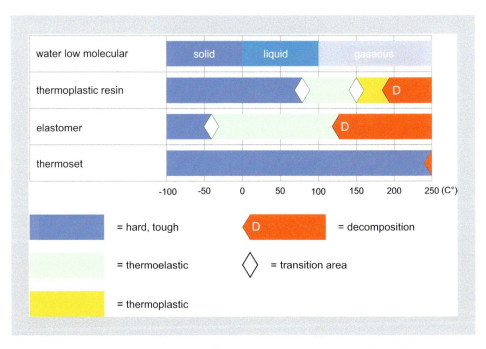

temperature ranges

Figure 4.4 The various physical states of water, elastomers, thermoplastics and thermosets at different temperatures

An example of an elastomer is natural rubber (see Figure 4.4). Natural rubber can be used at temperatures ranging from −40 to 130 °C (−40 to 266 °F).

If the plastic is highly cross-linked (thermoset), the rigidity of the plastic decreases only slightly, even in the softening range. Due to the many cross-linking sites between the individual molecular chains, the mobility of the macromolecules among each other is very limited. Like elastomers, thermosets cannot be melted. They are also destroyed when they exceed the decomposition temperature.

thermoset

The temperature ranges of a heat-resistant UP thermoset are shown in Figure 4.4 as an example of a thermoset. This thermoset can therefore be used at temperatures below 170 °C (338 °F).

temperature ranges

4.5 Performance Review – Lesson 4

No.	Question	Answer Choices
4.1	The elongation at maximum tension is referred to as _____.	tensile strength elongation at break
4.2	Tensile strength is a measure of the _____ of the plastic.	elasticity strength toughness
4.3	Elongation at break is a measure of the _____ of the plastic.	toughness tensile strength
4.4	The softening temperature range (ST) of semicrystalline thermoplastics is usually _____ room temperature.	below above
4.5	The CD is made of the _____ thermoplastic polycarbonate (PC) because it must have good transparent properties.	amorphous semicrystalline
4.6	The shear modulus of a cross-linked plastic is a measure of its _____.	rigidity strength toughness
4.7	Elastomers and thermosets do not melt because they are _____.	cross-linked amorphous
4.8	Natural rubber can be used in a range between approx. −40 °C (−40 °F) and approx. _____ °C (_____ °F).	+40 °C (104 °F) +130 °C (266 °F) +180 °C (356 °F)
4.9	In the summer, a CD _____ stored in a car (where temperatures can reach 80 °C) without fear of damage.	can be cannot be

5 Lesson
Time-Dependent Behavior of Plastics

Subject Area	Physics of Plastics
Key Questions	How does the strength of a loaded plastic change over time?
	What is meant by "creep" of a plastic?
	How do time and temperature dependence affect the strength of plastics and thus the use of this material?
Contents	5.1 Behavior of Plastics Under Load
	5.2 Effect of Time on Mechanical Behavior
	5.3 Recovery Behavior of Thermoplastics
	5.4 Dependence of Plastics on Temperature and Time
	5.5 Performance Review – Lesson 5
Prerequisite Knowledge	Classification of Plastics (Lesson 3)
	Deformation Behavior of Plastics (Lesson 4)

■ 5.1 Behavior of Plastics Under Load

In a tensile test, we simultaneously apply an identical load to a plastic and a metal sample. The samples expand, as shown in Figure 5.1 If the load were immediately removed from the samples, they would return to their original length.

tensile test

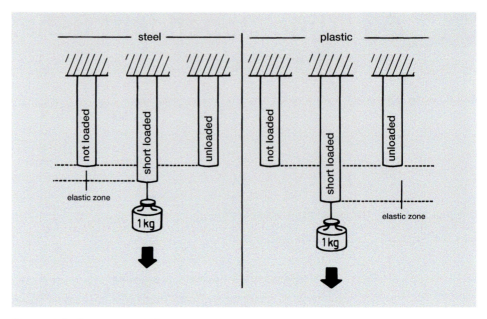

Figure 5.1 Deformation of the samples after short-term load application

stress
strain
modulus of elasticity E

However, we leave the load on them and measure the amount of elongation the specimens have undergone, called the "strain". The force with which the samples were loaded is divided by the cross-sectional area of the sample. This gives the stress applied to the specimen. If we now divide the stress by the strain, we get the modulus of elasticity for the specimen. This measure thus indicates the amount of strain a material undergoes when subjected to a given load, a measure of the material's strength. The higher the modulus of elasticity, the less strain occurs in a material under the same load, and the higher its rigidity. Table 5.1 now shows the modulus of elasticity E for various materials.

Table 5.1 Modulus of Elasticity for Various Materials

Material	Modulus of elasticity (N/mm²)
plastics	200–15 000
steel	210 000
aluminum	50 000

comparison: plastic vs. steel

It can be seen that the modulus of elasticity of steel, and thus its stiffness, is up to 1000 times higher than that of plastic. Therefore, in the specimen subjected to short-term loading (Figure 5.1), the change in length at the same load is smaller in the steel specimen than in the plastic specimen. As can be seen in Figure 5.1, it is evident for both groups of materials that they return to their original geometry after being subjected to a short-term load. Thus, steel and plastic do not differ in their behavior. However, while the modulus of elasticity from such a short-term test is of crucial importance for the design of technical metal parts, this is not the case for plastics. It only plays a minor role in the design of plastic parts, because it only allows a limited conclusion to be drawn about the strength of the plastic. This is because the stiffness of the plastic is time-dependent (see also Lesson 8).

■ 5.2 Effect of Time on Mechanical Behavior

Let's take another look at the two samples from earlier. The two specimens have now been under constant load for some time. If we measure the strain of the specimens again, we find that the metal specimen has the same strain as before (Figure 5.2). In the case of the plastic specimen, however, the strain has increased, although the load has remained constant. This is a characteristic of plastic which is called "creep".

creep

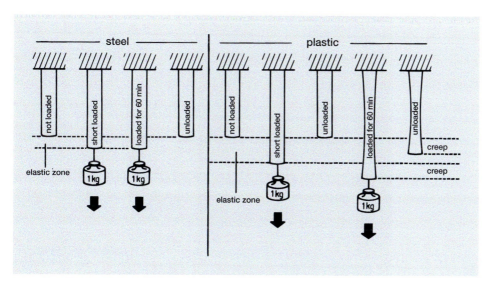

Figure 5.2 Deformation of specimens after extended load exposure

The figure shows that both specimens have the same length and dimensions when unloaded. If the same load is applied to both specimens, the steel specimen strains less than the plastic specimen due to the different elastic moduli of steel and plastic. While the steel specimen does not change even when subjected to extended loading, the behavior of the plastic differs significantly. It becomes longer over time, i. e., it "creeps". If both samples are unloaded again, both return to their original position under short-term loading, but only the steel body returns to its original position after extended loading. The plastic specimen lengthens accordingly by the amount of creep.

steel vs. plastic internal structure strain

The creep of the plastic can be explained from its internal structure. As shown, the plastic consists of tangled macromolecules held together by intermolecular forces. When the plastic is subjected to load, the tangles are the first to be strained.

slippage of the macromolecules

This strain also disappears if the plastic is immediately relieved and the load is relatively small. If the load is applied for an extended period, the intermolecular forces slowly dissipate. The macromolecules slide apart from each other. The plastic does not recover from the resulting strain, even when the load is removed.

viscoelasticity

The strain of the plastic is therefore partly elastic (tangle elongation) and partly plastic and viscous (slippage of the molecules). Therefore, the behavior of the plastic is called "viscoelastic". There is a model for this behavior, called the Maxwell model (Figure 5.3).

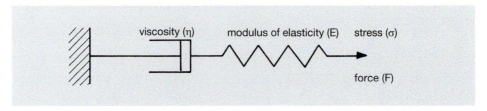

Figure 5.3 Maxwell model

Maxwell model

The model consists of a damper and a spring. When the model is loaded with a force, the spring spontaneously expands, while the damper does not yet react. Only if the load is maintained does the damper slowly begin to elongate. If the load is removed from the model, the strain of the spring will spontaneously recover, while the elongation of the damper will remain as viscous strain.

■ 5.3 Recovery Behavior of Plastics

macromolecules

As described so far, a tangle of macromolecules is somewhat strained under load, i.e., the macromolecules are stretched in length. If the plastic is immediately unloaded, the molecules return to their original position and the strain decreases.

recovery effect

A behavior of plastic based on the same principle is the recovery effect. We will look at a simple plastic tube in which the macromolecules exist in a tangled arrangement (Figure 5.4, position "a").

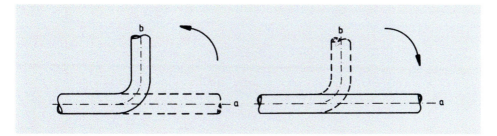

Figure 5.4 Recovery behavior of a plastic tube

We heat the tube until its temperature is within the thermoelastic range (above the softening temperature range ST, but below the flow temperature range FT). The tube can now be easily bent into a right angle (Figure 5.4, position "b"). This process of heating and shaping is also called "forming" in plastics processing.

thermoelastic range
forming

After forming, we cool the tube below the softening temperature (ST) range. The tube remains in deformation.

cooling

If we could examine the molecules at the point where the tube was bent, we would find that they are no longer tangled up there but stretched out. This is referred to as "orientation" in the plastic. But because the temperature is too low, they cannot move back to their original tangled shape. We say that the orientations are "frozen".

orientation

If we now reheat the deformed tube above the softening temperature (ST), the molecules move back to their initial position, thus pulling the tube back to its original straight shape (from position b to a in Figure 5.4). The orientations recover again. This process is called "recovery behavior".

recovery behavior

■ 5.4 Dependence of Plastics on Temperature and Time

As outlined, temperature and time have a decisive influence on the mechanical behavior of plastics. For this reason, the subsequent application temperature and load duration play a decisive role in the design of technical plastic parts. These factors are of less importance in the design of metal parts.

design

In order to give the designer a tool with which he can estimate these two influences, so-called "creep curves" are recorded from the individual plastics (Figure 5.5).

creep curves

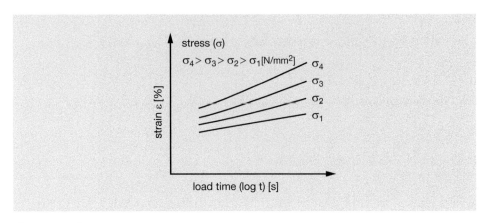

Figure 5.5 Creep curves

measurements	A force is applied to a plastic specimen with a defined cross-sectional area at a specified temperature, and the change in strain over time is measured. The tests are repeated with different force levels and at different temperatures.
diagrams	The measured values are plotted as curves in a diagram. The diagram (Figure 5.5) shows the strain of the samples as a function of loading duration, which is plotted here logarithmically. Each of the curves represents a particular load or stress (σ) and a particular temperature. The stress (σ_1) signifies the smallest stress applied and the stress (σ_4) the largest stress applied.
design	To simplify the diagrams, they often contain only curves that apply, for example, to a specific temperature. This makes it easy to understand how the curves change under different loads. The loads are given as stresses, i.e., as load per cross-section. This means that the values can be used to design parts of any desired cross-section.
	However, it is often more useful for the designer to have the information contained in a creep curve diagram in another form. Then they are converted to other diagram forms.
time-dependent creep diagram	One such other diagram is the time-dependent creep diagram. Here, the stress at a constant temperature and strain (ε) is plotted logarithmically as a function of time, as follows (Figure 5.6).

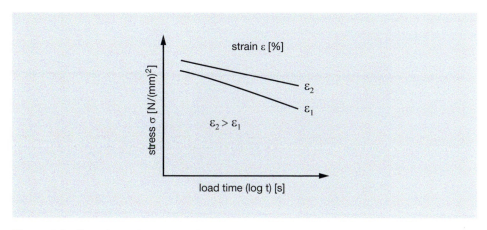

Figure 5.6 Time-dependent creep diagram

The time-dependent creep diagram provides a particularly good indication of the allowable stresses and thus allowable loads, considering that a certain strain of the part to be designed must not be exceeded.

Another diagram form is the isochronous stress-strain diagram. Here, the stress at a constant loading time (Greek: isos = equal; chronos = time) versus temperature is plotted as a function of strain (Figure 5.7).

stress-strain diagram

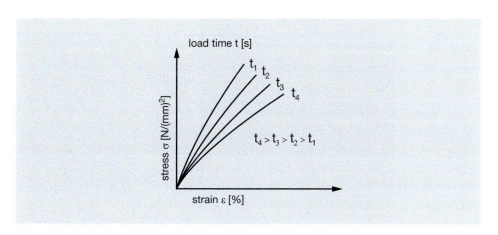

Figure 5.7 Isochronous stress-strain diagram

From this diagram, the area of the plastic in which the strain is linearly dependent on the stress can be seen particularly well.

Let us now look at how to interpret a creep diagram. Figure 5.8 shows a creep diagram for PMMA at 23 °C (73 °F). Let's assume that a part is subjected to a stress of 40 N/mm² (5800 psi) at 23 °C (73 °F). How long would it take for this part to break?

example PMMA

And applying the same assumptions for a stress of 50 N/mm² (7250 psi), how long would it take for it to break?

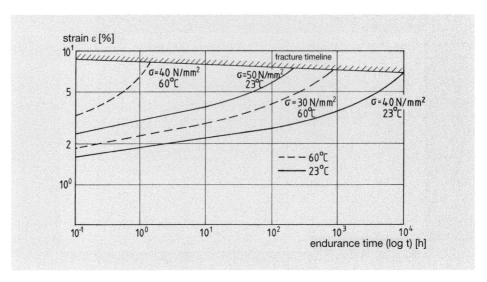

Figure 5.8 Temperature-dependent creep diagram for PMMA

result

For a stress of 40 N/mm² (5800 psi), this results in a lifetime of approximately 10^4 hours, which corresponds to a time of approx. 14 months. For a stress of 50 N/mm² (7250 psi), the lifetime is 2.6×10^2 hours, which equals approx. 11 days!

dimensioning

This demonstrates that the time-dependence of the behavior of a plastic must not be neglected in the design process in order to achieve appropriate dimensioning.

temperature dependence

Let us now examine PMMA at the same load of 40 N/mm² (5800 psi) but at different temperatures. As described before, at a temperature of 23 °C (73 °F), it takes approx. 14 months for PMMA to break. If we increase the temperature to 60 °C (140 °F), we find that it no longer takes 14 months, but only 1.5 hours (90 minutes) before the plastic breaks. The dependence on temperature is therefore even more drastic than the time dependence mentioned above.

metals vs. plastics

In summary, it can be concluded that, compared to most metals, plastics show a distinct dependence on the loading duration and the operating temperature. Both parameters, time dependence and temperature dependence, must be considered in the design of products made of plastics, whereby the influence of temperature is clearly more pronounced than the influence of time.

5.5 Performance Review – Lesson 5

No.	Question	Answer Choices
5.1	The modulus of elasticity is a measure of the _____ of a material.	strength rigidity plasticity
5.2	The modulus of elasticity is up to _____ times higher for steel than for plastics.	10 100 1000
5.3	The strain of a plastic _____ on the extent and duration of loading.	does not depend depends
5.4	The change in strain of a plastic under constant load is called _____.	slippage creep
5.5	The recovery effect can be seen when formed plastics are _____.	heated up cooled down
5.6	The existence of stretched frozen molecules is referred to as _____.	orientation stress
5.7	The significant dependence of the mechanical behavior of plastic on _____ must be considered in the design process.	time and temperature temperature time
5.8	The designer can use diagrams such as creep curves, _____ or the isochronous stress-strain diagram as aids in design.	time-dependent creep diagrams fatigue strength chart
5.9	In the case of plastics, the influence of temperature is _____ pronounced than the influence of time.	less more

6 Lesson
Physical Properties

Subject Area	Physics of Plastics
Key Questions	How does the density of plastics compare with that of metals?
	How well do plastics conduct heat?
	How well do plastics conduct electricity?
	What are the optical properties of plastics?
Contents	6.1 Density
	6.2 Thermal Conductivity
	6.3 Electrical Conductivity
	6.4 Transparency
	6.5 Material Characteristics of Plastics
	6.6 Performance Review – Lesson 6
Prerequisite Knowledge	Plastics Fundamentals (Lesson 1)

■ 6.1 Density

Low density ρ is a characteristic feature of plastics compared to other materials (Table 6.1). The density of plastics ranges from 0.9 g/cm³ to 2.3 g/cm³. Examples of low density plastics are the commodity plastics polyethylene (PE) and polypropylene (PP). Both materials have a lower density than water. Thus, they float in water. For this reason, it is also possible, for example, to separate these two plastics from heavier plastics due to their greater buoyancy in water. Most plastics are within the density range of 1 g/cm³ to 2 g/cm³. And only a few have densities exceeding 2 g/cm³, such as polytetrafluoroethylene (PTFE).

density range

6 Physical Properties

Table 6.1 Density of Various Materials

Material	Density ρ (g/cm³)
plastics	0.9–2.3
▪ PE	0.9–1.0
▪ PP	0.9–1.0
▪ PC	1.0–1.2
▪ PA	1.0–1.2
▪ PVC	1.2–1.4
▪ PTFE	>1.8
steel	7.8
aluminum	2.7
wood	0.2–0.95
water	1.0

explanation

The density of other materials is in some cases several times higher. For instance, the density of aluminum is about 2.7 g/cm³ and of steel 7.8 g/cm³. The higher density of other materials is due to two factors:

- The individual atoms (aluminum, iron) are heavier than the atoms from which plastics are composed, i.e., carbon, nitrogen, oxygen, or hydrogen.
- The average distance between the atoms is partly larger in plastics than in metals.

■ 6.2 Thermal Conductivity

thermal conductivity

The thermal conductivity of a material is a measure of how well it can transport heat. For plastic, the thermal conductivity is in the range 0.15 to 0.5 W/(m K). This is a very poor value. Table 6.2 lists the thermal conductivities of other materials in comparison to plastic. Metal, for instance, has a value up to 2000 times higher. It conducts heat very well. Air, on the other hand, conducts heat 10 times less efficiently than plastic.

Table 6.2 Thermal Conductivity of Different Materials

Material	Thermal conductivity λ (W/mK)
plastic	0.15–0.5
▪ PE	0.32–0.4
▪ PA	0.23–0.29
steel	17–50

Material	Thermal conductivity λ (W/mK)
aluminum	211
copper	370–390
air	0.05

One reason for the low thermal conductivity of plastic is the lack of freely moving electrons in the material. Since metals have such electrons, they are good conductors of both electric current and heat. *explanation*

One disadvantage of the poor thermal conductivity becomes apparent when processing plastics. The heat required for processing can only be slowly introduced into the plastic and is also difficult to dissipate at the end of processing. *processing*

What proves to be a disadvantage in processing, however, is often an advantage in everyday use. For example, plastics are used as pot handles because they do not get hot as quickly as metal when the pots are heated, so you can take the pot off the stove without burning your fingers. Plastics are also used as insulating materials in the construction industry. Since air, as described earlier, conducts heat even less, air is "admixed" to the plastic. This results in a foamed plastic that has the average value of the two thermal conductivities. On the other hand, you can add metallic fillers to the plastic to increase its thermal conductivity. *applications*

■ 6.3 Electrical Conductivity

A measure of how well a material can conduct electricity is its electrical conductivity. Plastics generally conduct electric current very little. They have high resistances and thus low conductivity compared to other materials (Table 6.3). The electrical resistance of plastics is temperature dependent. It decreases with increasing temperature; the plastic then conducts better. *electrical conductivity*

Table 6.3 Electrical Conductivity

Material	Electrical conductivity Σ (m/Ohm mm²)
PVC	10^{-15} (up to approx. 60 °C)
steel	5.6
aluminum	38.5
copper	58.5

One reason for the low electrical conductivity of plastics is the absence of the free electrons that occur in metals. *explanation*

conductivity increase

If a better conductivity of the plastic is requested, metal powder can be incorporated. The resulting effect on the electrical resistance R of the plastic can be seen in Figure 6.1.

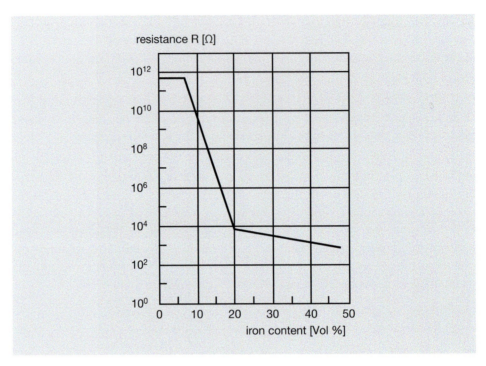

Figure 6.1 Resistance of a metal-powder-filled plastic

insulation

As can be seen, the resistance decreases by a factor of 10,000,000 (10 million!) when 20% by volume of metal powder is admixed. Because of their very high electrical resistance, plastics are preferred for insulating electrical devices and cables.

■ 6.4 Transparency

transmittance

Light transmission or transmittance is the ratio of the strength of the light transmitted without deflection to the strength of the incoming light. Amorphous thermoplastics such as PC, PMMA, PVC and UP resin do not differ significantly from window glass in their light transmission. Their light transmission is about 90% (Table 6.4). This is equivalent to a transmittance of 0.9%, which means that a factor of 0.1 or 10% of the light is lost through reflection and absorption.

Table 6.4 Transparency

Material	Transparency (%)
PC	72–89
PMMA	92
window glass	90

One disadvantage of plastics, however, is that environmental influences, such as weathering or thermal cycling, can cause hazing and thus a decline in light transmission.

environmental impacts

Optical data carriers (CD, CD-ROM, DVD) are made of the amorphous thermoplastic polycarbonate (PC), which, in addition to other worthy properties, is characterized by its good light transmission, which comes close to that of glass.

CD

■ 6.5 Material Characteristics of Plastics

The physical properties of plastics presented here are not all the properties that a plastic possesses. The various properties can be classified into classes, such as the mechanical property class or the thermal property class. The various properties of each class usually include several physical measured values that describe the property of the plastic. Based on these measured values, a designer or planner can select the plastic that meets specific requirements.

property categories

To manufacture a CD through injection molding, for example, an easy-flowing plastic is required so that the fine indentations containing the data information (music, images, etc.) are completely mapped. This ensures that no "yes-no" information gets lost.

databases

This means that a plastic with an extremely low viscosity is required for the manufacturing process. In addition, the plastic must still be very transparent, so that the data information can be read. Thus, an amorphous thermoplastic must be selected. Of course, other properties are also important and must be included in the "search profile".

low viscosity CD/DVD

In former times, spreadsheets and manuals were used to determine the material data for a plastic possessing certain material properties. Nowadays, raw material manufacturers make all relevant material data for their plastics accessible in a database. The engineer or expert can then easily and conveniently search for the right material using a specific search profile that replicates the material properties for a particular application and compare it with data from other manufacturers. One database widely used around the world is CAMPUS, a data sheet from which is shown in Figure 6.2 using ABS material as an example.

search profile

ABS

This ABS (acrylonitrile butadiene styrene) has the brand name "Novodur" (which is protected by law) and bears the company designation "P2H-AT". In addition to the material value and the unit, the test standard, according to which the values were determined, is also indicated here, so that they can also be compared. For example, the density of this plastic is 1050 kg/m^3, which is equivalent to 1.05 kg/dm^3. This plastic is therefore slightly heavier than water, which has the standard value of 1.00 kg/dm^3.

VDA CAMPUS® Datasheet
Novodur® P2H-AT - ABS
Styrolution

Physical properties	I	M	E[1]	Value	Unit	Test Standard
Melt volume-flow rate, MVR	X	X	X	37	cm³/10min	ISO 1133
Temperature	X	X	X	220	°C	ISO 1133
Load	X	X	X	10	kg	ISO 1133
Viscosity number	X	X	X	-	cm³/g	ISO 307, 1157, 1628
Molding shrinkage, parallel	X	X	X	-	%	ISO 294-4, 2577
Molding shrinkage, normal	X	X	X	-	%	ISO 294-4, 2577
Humidity absorption	X	X	X	-	%	Sim. to ISO 62
Water absorption	X	X	X	-	%	Sim. to ISO 62
Density	X	X	X	1050	kg/m³	ISO 1183
Type and amount of reinforcement	X	X	X	-	-	ISO 3451-1
Mechanical properties	**I**	**M**	**E[1]**	**Value**	**Unit**	**Test Standard**
Tensile Modulus	X	X	X	2500	MPa	ISO 527-1/-2
Yield stress	X	X	X	44	MPa	ISO 527-1/-2
Stress at break	X	X	X	*	MPa	ISO 527-1/-2
Yield strain	X	X	X	2.1	%	ISO 527-1/-2
Strain at break	X	X	X	*	%	ISO 527-1/-2
Charpy impact strength, +23°C	X	X	X	100	kJ/m²	ISO 179/1eU
Charpy notched impact strength, +23°C	X	X	X	16	kJ/m²	ISO 179/1eA
Charpy impact strength, -30°C	X	X	X	80	kJ/m²	ISO 179/1eU
Charpy notched impact strength, -30°C	X	X	X	7	kJ/m²	ISO 179/1eA
Puncture test - ductile/brittle transition temperature	X		X	-	°C	ISO 6603-2
Thermal properties	**I**	**M**	**E[1]**	**Value**	**Unit**	**Test Standard**
Melting temperature, 10°C/min	X	X	X	*	°C	ISO 11357-1/-3
Temp. of deflection under load, 1.80 MPa	X	X	X	93	°C	ISO 75-1/-2
Temp. of deflection under load, 0.45 MPa	X	X	X	97	°C	ISO 75-1/-2
Temp. of deflection under load, 8.00 MPa	X	X	X	*	°C	ISO 75-1/-2
Vicat softening temperature, 50°C/h 50N	X	X	X	98	°C	ISO 306
Coeff. of linear therm. expansion -40°C to +100°C, parallel	X	X	X	-	E-6/K	ISO 11359-1/-2
Coeff. of linear therm. expansion -40°C to +100°C, normal	X	X	X	-	E-6/K	ISO 11359-1/-2
Burning rate, Thickness 1 mm	X			-	mm/min	ISO 3795 (FMVSS 302)
Burning Behav. at 1.5 mm nom. thickn.		X	X	HB	class	IEC 60695-11-10
Emission / Odor	**I**	**M**	**E[1]**	**Value**	**Unit**	**Test Standard**
Emission of organic compounds	X			-	µgC/g	VDA 277
Thermal desorption analysis of organic emissions	X			-	µg/g	VDA 278
Odor test	X	X[2]		-	class	VDA 270
Long term / Aging	**I**	**M**	**E[1]**	**Value**	**Unit**	**Test Standard**
Thermal stability in air, Charpy at 50% decrease, 3000h	X	X	X	-	°C	DIN/IEC 60216-1
Test specimen				-	-	
Weather stability, ISO 4892-2, Method A	**I**	**M**	**E[1]**	**Value**	**Unit**	**Test Standard**
Weather stability delta l			X	-	-	DIN 53236
Weather stability delta a			X	-	-	DIN 53236
Weather stability delta b			X	-	-	DIN 53236
Weather stability delta E			X	-	-	DIN 53236
Weather stability grey scale			X	-	-	ISO 105-A02

[1] I=Interior parts, M=Parts in motor compartment, E=Exterior parts
[2] air-ducting parts with contact to interior

Datasheet according to an agreement between VDA (Association of the Automotive Industry) and CAMPUS®
All data is subject to the producer's disclaimer.
http://www.campusplastics.com - Styrolution - 2013-02-28

Figure 6.2 Extract from a data sheet for an ABS plastic (*http://www.campusplastics.com*)

However, plastics databases offer many more possibilities beyond this data sheet, diagrams such as complex correlations in diagram form. Figure 6.3 and Figure 6.4 show isochronous stress-strain diagrams for two different temperature ranges 23 °C (73 °F) and 40 °C (104 °F) as a function of different loading periods (1 h to 10 000 h).

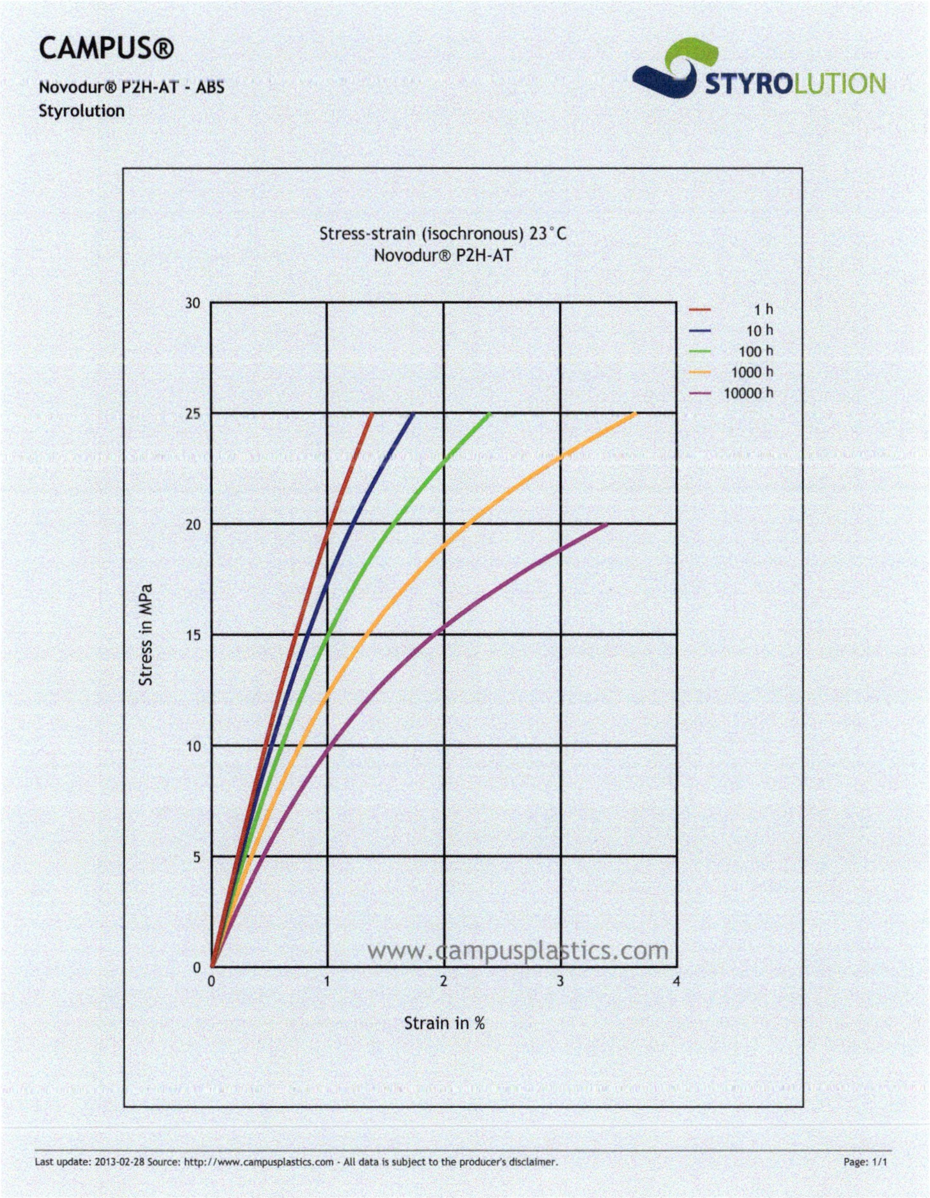

Figure 6.3 Isochronous stress-strain diagram for ABS at 23 °C (73 °F) (*http://www.campusplastics.com*)

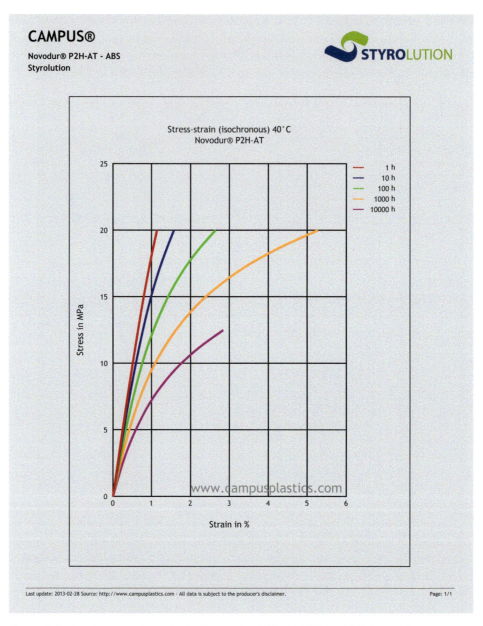

Figure 6.4 Isochronous stress-strain diagram for ABS at 40 °C (104 °F) (*http://www.campusplastics.com*)

ABS

As can be seen for the ABS material, for a loading duration of 1000 hours' continuous exposure to a temperature of 40 °C (104 °F), the strain is 5.2%, at a stress of 20 MPa. At a room temperature of 23 °C (73 °F), the strain value is 2.3% for the same time period of 1000 h and the same stress of 20 MPa. Under the temperature influence of only plus 17 K, the plastic part experiences more than twice as much strain. This is a major difference compared with metallic materials. As long as they are not loaded beyond the modulus of elasticity, the duration of loading does not play a role in the strain. In the first case, a plastic component with a length of 1 m, for example, would have become 23 mm longer at a temperature of 23 °C after 1000 h (approx. 40 days) of loading and subject to a stress of 20 MPa. This linear expansion would not be reversible. If the temperature is increased to 40 °C during the entire loading period and under the same load, the linear expansion would have increased to 52 mm.

■ 6.6 Performance Review – Lesson 6

No.	Question	Answer Choices
6.1	Plastics generally are _____ than metals.	lighter heavier
6.2	The density of steel is 7.8 g/cm³. The density of plastics is in the range _____ g/cm³.	0.5 to 0.8 0.9 to 2.3 2.5 to 5.0
6.3	Metals are up to _____ times more thermally conductive than plastics.	20 200 2000
6.4	The poor electrical conductivity of plastics can be improved with additives such as_____.	chalk flour metal powder glass splinters
6.5	The light transmission of amorphous thermoplastics is _____ that of glass.	greater than less than roughly equal to
6.6	A CD is made of the amorphous plastic PC because of its good _____.	thermal conductivity light transmission density
6.7	A loading duration of 1000 h corresponds approximately to a period of _____ days.	82 41 123
6.8	At a strain of 2.4% and a loading duration of 1000 h as well as a defined stress, the change in expansion of a plastic rod with an initial length of 500 mm is about _____ mm.	24 48 12

7 Lesson
Fundamentals of Rheology

Subject Area	Plastics Fundamentals
Key Questions	What is meant by rheology?
	What is meant by shear stress?
	What is meant by shear rate?
	What is meant by viscosity?
	Which flow behavior do plastic melts exhibit?
	How can flow behavior be determined?
Contents	7.1 Fundamentals
	7.2 Flow and Viscosity Curves
	7.3 Flow Behavior of Plastic Melts
	7.4 Melt Flow Index MFI
	7.5 Performance Review – Lesson 7
Prerequisite Knowledge	Plastics Fundamentals (Lesson 1)
	Raw Materials and Polymer Synthesis (Lesson 2)
	Classification of Plastics (Lesson 3)

■ 7.1 Fundamentals

Rheology (from the Greek for "study of flow") is a branch of physics and generally describes the flow behavior of substances (solids, liquids, gases) under the effect of external forces. For plastics, rheology is of particular importance because most plastics are processed by melting, thus produced in liquid form as a fluid (liquid) under heat and subsequent cooling. Since plastics are poor conductors of heat, liquefaction of the solid plastic (pellets) and transport of the liquid plastic for the

rheology
flow behavior

normal stress
shear stress

strain

manufacturing of products are core tasks of plastics processing and also key requirements for plastics processing machines.

Solids can be strained by tensile loads, i.e., by the application of a normal force. However, they can also be deformed by a shear stress. Liquids, such as water, can only be subjected to shear stresses. Stresses in solids (load/surface) cause a change in shape.

This change in shape is called "strain" and corresponds to the change in length relative to the initial length. Shear stresses, on the other hand, correspond to angular changes. A shear stress applied to a theoretical fluid element causes deformation from a rectangle to a parallelogram (Figure 7.1). The right angle changes by the angle α, which describes the degree of deformation.

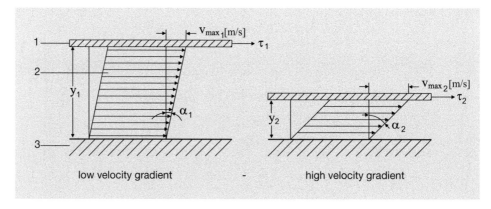

Figure 7.1 Schematic representation of shear flow with a moving plate (two-plate model)

Shear Stress

shear stress

As shown in Figure 7.1, the plate is moved to the right with a tangential force, causing the fluid to flow. The ratio of the force F in relation to the plate area A is called the "shear stress". The symbol for shear stress is the Greek letter τ (pronounced: "tao") and its physical unit is the Pascal [Pa].

$$\tau = \frac{F}{A} = \frac{N[\text{Newton}]}{m^2} = Pa[\text{Pascal}] \tag{7.1}$$

τ shear stress F force A area

shear rate

The shear stress induces flow in a fluid. Looking at the two-plate model (Figure 7.1), we can observe a velocity drop across the individual fluid layers, from a maximum value directly at the moving plate to zero at the lower interface. This velocity drop is called the shear rate and is symbolized by $\dot{\gamma}$ (read: "gamma point"). The shear rate defines the difference in flow velocity between two fluid layers and its unit is 1/s.

$$\dot{\gamma} = \frac{dv}{dy} = \frac{m/s}{m} = \frac{1}{s} = s^{-1} \qquad (7.2)$$

$\dot{\gamma}$ shear rate dv velocity dy height (y-axis)

For the processing of plastics, the shear rate is mainly determined by the volumetric flow rate and the geometry of the flow channel. The shear rate increases when the volume flow is increased or when the flow channel is decreased.

Viscosity

Another important property for characterizing flow properties is viscosity. The viscosity describes the flow resistance of a fluid during shear. If we consider our two-plate model (Figure 7.1), viscosity is defined as the ratio of shear stress and shear rate. The symbol for viscosity is the Greek letter η (pronounced: "eta") and the unit is Pa s (Pascal seconds).

$$\eta = \frac{\tau}{\dot{\gamma}} = \frac{\text{Pa}[\text{Pascal}]}{1/s} = \text{Pa s}[\text{Pascal seconds}] \qquad (7.3)$$

η viscosity τ shear stress $\dot{\gamma}$ shear rate

Table 7.1 shows some typical viscosities of well-known materials used in everyday life.

Table 7.1 Typical Viscosity Values of Some Substances at 20 °C (68 °F)

Material	Viscosity η (Pa s)	Material	Viscosity η (Pa s)
air	0.00001	coffee cream	10
water	0.001	honey	10^4
olive oil	0.1	plastic melt*	100–1 000 000
glycerine	1.0	pitch	10^9
grape juice	2–5	glass	10^{21}

* at processing temperature

Water, for example, flows more easily than honey at room temperature and therefore has a lower viscosity than honey. Honey has a high viscosity, so it flows more slowly. It is "viscous".

7.2 Flow and Viscosity Curves

flow curve

The flow behavior of liquids can be represented by means of diagrams. A flow curve describes the correlation between shear stress and shear rate. Here, the values of the shear rate are plotted on the x-axis and those of the shear stress on the y-axis in diagram form.

viscosity curve

Another common representation for the characterization of flow behavior is the viscosity curve. The viscosity curve represents the dependence of viscosity on shear rate. Usually, flow curves are generated first for viscosity measurements, which are then converted to viscosity curves. Figure 7.2 shows the corresponding flow and viscosity curves for a "Newtonian fluid".

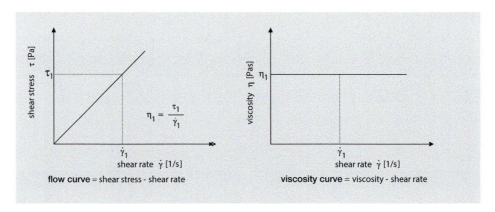

Figure 7.2 Flow curve and viscosity curve of a Newtonian fluid

The classic example of a Newtonian fluid is water.

7.3 Flow Behavior of Plastic Melts

Newtonian and Non-Newtonian Fluids

Newtonian fluid

In a Newtonian fluid, the shear stress is proportional to the shear rate. In the flow curve, all ratios of the two values τ and $\dot{\gamma}$ are constant. This means that the viscosity η of a Newtonian fluid is independent of the shear rate. All fluids that have this characteristic are called Newtonian. Examples of Newtonian fluids are water, mineral oils, or bitumen.

non-Newtonian fluid

Fluids that exhibit a different flow behavior are called "non-Newtonian" and occur much more often in practice. Plastic melts, i.e., molten plastics, belong to the group of non-Newtonian fluids, although they also exhibit Newtonian ranges in their flow behavior.

Shear Thinning Fluids

As can be seen in Table 7.1, plastic melts have a much higher viscosity than water, for example. The difference in magnitude is about five powers of 10. For high viscosities, the processing machines therefore also require high torque in order to be able to extrude the melt through the molding dies.

polymer melts

As we already learned in lesson 1, a plastic consists of many chain molecules. The viscosity of a plastic depends, among other things, on the length of these chain molecules. Long chains of molecules can become more tangled with each other, making them harder to shear off than short-chain molecules. The length of molecules is described by the "molecular weight". Therefore, there is also a correlation between the molecular weight and the viscosity of a plastic. A high molecular weight also implies a high viscosity.

molecular weight

To properly understand the plastics processing procedure, it is therefore important to know how the melt behaves under the given conditions. When plastics are processed on a screw machine, the material is sheared. This means that individual layers within the melt flow at a different rate. The degree of shear occurring in a fluid is called the shear rate.

plastics processing procedure

Plastic melts exhibit non-Newtonian flow behavior, or more precisely, they belong to the group of shear-thinning fluids. Figure 7.3 shows the flow curve and viscosity curve of a thermoplastic.

flow behavior

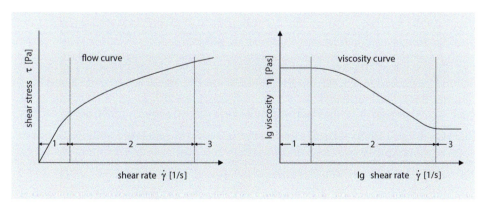

Figure 7.3 Flow curve and viscosity curve of shear-thinning fluids (qualitative illustration)

At low shear rates, shear-thinning fluids behave similarly to Newtonian fluids. The viscosity does not depend on the shear.

shear rate sections

section 1	The viscosity at extremely low shear rates is called "zero viscosity" or "first Newtonian range" (Figure 7.3 – section 1).
section 2	At higher shear rates (Figure 7.3 – section 2), we can observe the rheological phenomenon that typifies shear thinning. The shear of the melt causes alignment of the filament molecules in the direction of flow and thus a disentanglement or elimination of the tangled structure. The more the molecules align, the easier it is for the molecular chains to slide off one another. Accordingly, the flow resistance decreases with increasing shear, i.e., the plastic flows more easily. In the opposite direction, the viscosity values increase again under decreasing shear.
section 3	Beyond a certain shear rate, the molecules have reached a maximum alignment. In this range, a further increase in shear will not cause a further decrease in viscosity (Figure 7.3 – section 3). This region is called the "second Newtonian range".
effect of temperature	The viscosity of a fluid is also dependent on other factors. In the case of plastic melts, for example, the shear rate and the temperature have a major effect on the viscosity. An increase in temperature causes a decrease in viscosity and results in the plastic melt flowing more easily.

■ 7.4 Melt Flow Index (MFI)

melt flow indexer	A widely used method for characterizing the flow properties of plastic melts is measurement with the melt flow indexer. Both the measuring apparatus and the procedure (Figure 7.4) are recognized and standardized worldwide.
melt flow index MFI	The measured value is called the "melt flow index" (MFI) and can be found in all the data sheets supplied by raw material suppliers.

The measuring device consists of a capillary, a piston, and a storage cylinder. The material is heated in the storage cylinder so that it melts. The piston is loaded with weights from the top and thus forces melt through the capillary. The emerging melt is cut off and weighed at intervals of 10 minutes. The melt flow index indicates the weight (in grams) of plastic melt pushed through the capillary over a defined period (10 minutes). The dimensions of the capillary, the cylinder, the piston, and the weights are standardized (Figure 7.4).

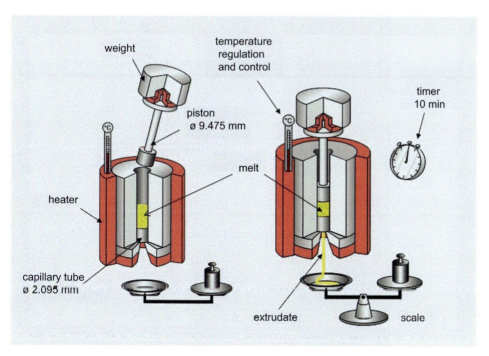

Figure 7.4 Schematic illustration for the determination of melt viscosity

The melt flow may also be stated as melt flow rate (MFR), which is equivalent to the melt flow index (MFI).

MFR – melt flow rate/ MFI – melt flow index

A common commercial polyethylene will be used as a typical example. This PE has an MFR value of 0.4 to 0.7 (g/10 min) under the test conditions MFR/190/5. This indicates that the test is carried out at a test temperature of 190 °C with a mass of 5 kg.

example: PE

The material PE-LD is used to produce carrier bags and garbage bags because it can be manufactured at low cost. For example, films made of this material can reach a maximum strain (vertical) of 500% before rupturing and this corresponds to a factor of 5. The thickness of such films for plastic carrier bags is between 60 and 120 microns. For comparison, a human hair has a thickness of 50–70 microns, which is about the same thickness as a carrier bag. Smaller waste bags can be made even significantly thinner.

example: plastic carrier bag

Due to the extreme simplicity in determining the flowability of a plastic, this method is used in production control, incoming goods inspections, and failure analysis or to deal with customer complaints. It is a standard tool in quality assurance.

quality assurance using MFR

7.5 Performance Review – Lesson 7

No.	Question	Answer Choices
7.1	The shear stress is the quotient of force and _____.	shear surface velocity
7.2	The velocity gradient of fluid layers is called _____.	viscosity shear rate
7.3	The viscosity of a Newtonian fluid _____ the shear rate.	depends on is independent of
7.4	The shearing of polymers causes _____ of the molecular chains.	alignment entanglement
7.5	The viscosity describes the _____ of a fluid.	flow resistance velocity drop
7.6	For shear-thinning fluids, raising the shear rate causes the viscosity to _____.	increase decrease
7.7	Water shows a typical _____ flow behavior.	shear thinning Newtonian
7.8	An increase in temperature causes _____ in the viscosity.	an increase a decrease
7.9	The flow behavior of plastics shows a _____ characteristic.	Newtonian shear thinning
7.10	The viscosity at low shear rates is called _____.	zero viscosity base viscosity
7.11	The melt flow index MFI indicates how many grams of plastic are forced through the capillary (narrow opening) of the melt flow index tester within _____ minutes.	10 20 30

Lesson 8
Plastic Applications

Subject Area	From Plastic to Product
Key Questions	What must be considered when selecting plastics for specific applications?
	How do specifications influence the choice of materials?
	What is the connection between components/products and processing methods?
	What significance does recycling have for the choice of material?
Contents	8.1 Fundamentals
	8.2 Requirements Criteria – Material Selection – Manufacturing Processes
	8.3 Examples for Plastic Applications
	8.4 Performance Review – Lesson 8
Prerequisite Knowledge	Plastics Fundamentals (Lesson 1)
	Classification of Plastics (Lesson 3)
	Deformation Behavior of Plastics (Lesson 4)
	Time-Dependent Behavior of Plastics (Lesson 5)
	Physical Properties (Lesson 6)

8.1 Fundamentals

plastics

This lesson is about developing a basic understanding of the correlation between the requirement profiles resulting from the use of products in their typical environment and finding a solution for this by means of suitable plastics. In this context, inexpensive, simple plastics (commodities) are discussed, together with (expensive) engineering plastics. Some of the selected examples are also covered in other lessons in the book, such as manufacturing processes, materials, as well as recycling and the waste problem. The idea is to provide an holistic understanding of the use of plastics.

integrating functions
new solutions

This lesson is also intended to draw attention to the fact that the quality of components made of plastics depends very heavily on the processing method chosen. In addition to the compact disc (CD) previously described in the book, other examples will be explained, including cases that show the possibilities of functional integration of plastics. In the case of injection molding, for example, the advantage of many plastics results in a finished part having a complex geometry and various integrated functions. These include snap-fits, snap closures, spring elements and film hinges, as well as bearings that function without lubricants – all manufactured in a single operation. Plastic as a material can be used to perform many different functions without the need for additional manufacturing steps. This is one of the success factors behind the steady growth and global spread of this material class.

8.2 Requirements Criteria – Material Selection – Manufacturing Processes

functionality
specification profile

All products and components must meet certain requirements, which the customer or buyer ultimately defines through his purchase decision. Of course, legal requirements also play a role, and the design is important, which likewise can be a requirement criterion. The fundamental factor in most cases is therefore the functionality that a product must provide and guarantee. Of course, this must be seen in relation to the price that a customer is willing to pay for the product. Table 8.1 lists examples of various components and products and their specifications, along with the chosen material, the manufacturing process, and the possibility of recycling. The listed products and components reveal a unique feature of plastics. There is no standard manufacturing process, but for each component or product and chosen material, consideration must be given to the processing technique and, if necessary, further treatment or machining of the products.

Table 8.1 Product, Specification Criteria, Material Selection and Manufacturing Process

No.	Component/product	Specification criteria	Material	Processing/machining technique	Recycling
1	CD/DVD/HD-DVD (Blu-ray)	flowability, transparency, breaking strength	PC	injection molding/injection compression molding	yes
2	window profiles	environmental resistance, UV resistance, weldability, heat resistance	PVC	extrusion	yes
3	windows	environmental resistance, UV resistance, weldability, heat resistance	PVC, metal	welding/cutting	yes
4	translucent twin-wall sheets	transparency, insulation properties, scratch resistance, impact resistance	PMMA	extrusion	yes
5	water bottles	transparency, UV resistance, hygienically safe	PC	injection molding	yes
6	garbage bins	chemical resistance, UV resistance	PE	injection molding	yes
7	trash bags	tear resistance, elasticity, cost, moisture resistance	PE	blown film extrusion	yes
8	shopping bags	tear resistance, elasticity, cost, moisture resistance	PE	blown film extrusion	yes
9	DVD boxes	flexural strength, notch effect	PP	injection molding (integral hinge)	yes
10	DVD boxes	flexural strength, notch effect	PP/POM	injection molding (snap hinge)	yes
11	piping	environmental resistance, weldability	PVC/PE	extrusion + welding	yes
12	fuel tanks	gasoline/diesel resistance, diffusion tightness	PE	extrusion blow molding	limited

Table 8.1 Product, Specification Criteria, Material Selection and Manufacturing Process (*continued*)

No.	Component/product	Specification criteria	Material	Processing/machining technique	Recycling
13	tennis rackets	flexural strength, lightweight	CFRP	hand lay-up	limited
14	shrink sleeves	memory effect	PE/PP	extrusion	yes
15	beverage bottles	transparency/UV resistance	PET	injection blow molding	yes

This lesson does not explain design principles or provide calculation examples. It is primarily concerned with a basic understanding of the selection of a suitable plastic for specific components and products. In particular, the relationship between the requirements specification, material selection, processing methods or machining processes and the aspect of recycling is explained. Some of the selected examples can be found in the lessons of this book under the respective contents dealt with there (materials, processing methods, recycling, etc). They are complemented by two additional examples which illustrate certain particularities of plastics compared with other materials.

plastic vs. paper

Plastics are extremely versatile and meet so many requirements that other materials are less able to fulfill. For example, carrier bags made of PE or PP have the undisputed advantage over paper of being significantly more tear-resistant and thus having a higher load bearing capacity. In addition, plastics do not lose their load bearing capacity when exposed to moisture (e.g., rain), as is the case for paper carrier bags. However, plastic carrier bags have the disadvantage over paper carrier bags of being much more difficult to dispose of, as they have long expiration times and do not biodegrade quickly, unless they are made of biopolymers. Suitable collection systems and recycling processes are therefore crucial, and the carrier bags must not end up in the environment or the sea under any circumstances.

plastic vs. metal

Materials are in competition with each other. For thermoplastics, many rules must be observed because of the greater time and temperature dependence of the material properties described above. Thus, in terms of fatigue strength, plastics cannot compete with steel and aluminum in many applications that are exposed to elevated temperatures.

Thermosets, on the other hand, can be superior to aluminum and steel in terms of their load-bearing capacity. Fiber-reinforced plastics (FRP), for example, are commonly used in aircraft construction and sports, such as in Formula 1 for the vehicle chassis and for drive shafts. Other products such as tennis rackets and modern lightweight bicycles are also made from FRP.

The basis of any design is always to select the suitable material for the respective application. However, not only strengths, stiffness, or surface properties play an important role, but also the different plastic processing methods and thus a process-compliant design. Particularly in the production of thermosets and thermoplastics, there are major differences. In any production of products, costs naturally play a significant role and so do the material costs. Especially in mass production (plastic pipes, plastic profiles, carrier bags, garbage bags, plastic beverage bottles, films in agriculture, etc.), material costs are a substantial cost factor and can easily account for 50% and more of total costs.

design
material costs
mass production

These commodities (high-volume products) are manufactured under a high degree of automation, for example, injection molding or extrusion. In automated manufacturing processes, personnel are usually employed for set-up and start-up processes as well as for machine and system monitoring. This contrasts with the situation in assembly plants, such as those in the automotive industry.

manufacturing processes:
injection molding
extrusion
laminating
welding
adhesive bonding
additive manufacturing

While thermoplastics are used particularly in mass production, thermosets are more common in small batches, where labor costs play a greater role in relation to material costs. Plastics are processed and machined by the classic methods of extrusion (profile extrusion, extrusion blow molding, blown or flat film extrusion), injection molding, laminating, welding, and bonding, each of which is covered by a lesson in this book. Also covered are the processes of additive manufacturing, which open up entirely new options for manufacturing with regard to production time and costs.

The example of injection molding will be used to show process characteristics that can influence component quality and must therefore be taken into account for the selection of a suitable plastics processing method. So-called "weld lines" (also known as "flow lines") always occur when plastic flow fronts meet (Figure 8.1). The smaller the temperature difference when partial melts flow together, the more homogeneously they converge and the lower is the effect on component strength.

injection molding
weld lines

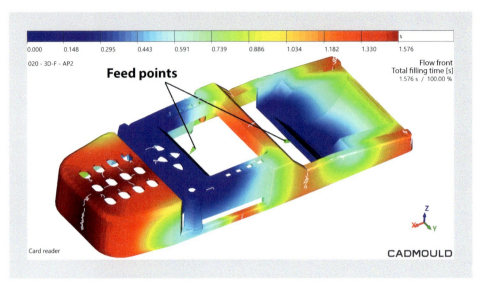

Figure 8.1 Injection molded part (magnetic card reader) with weld line and feed points

Visible weld lines, for example, represent a visual reduction in quality in the case of high-gloss cell phone shells, and they may also affect mechanical strength.

■ 8.3 Examples of Plastic Applications

examples

Some examples from Table 8.1 will be discussed in detail here. They will focus on the connection between the specification criteria, the proper choice of material, the right manufacturing process, and the consideration of recycling at the end of the product's service life.

Translucent Twin-Wall Sheets

twin-wall sheets
performance criteria

Twin-wall sheets are used, for example, as glass substitutes for patio roofs or in conservatories. The name twin-wall sheet is derived from the special profile of these sheets, which are provided with a bridge between the upper and lower layers for strength reasons (Figure 8.2). Compared with the competitive product glass, they have similar transparency properties, but because of their double-webbed design they have a higher load-bearing capacity and are also significantly more elastic and resistant, especially to breaking loads.

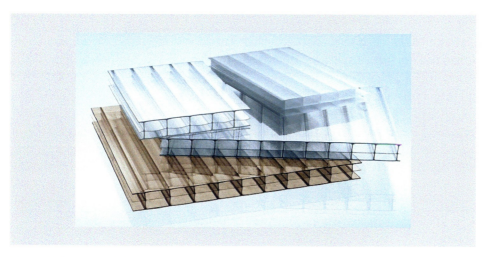

Figure 8.2 Twin-wall sheets

Twin-wall sheets are made from the plastics polymethyl methacrylate (PMMA) and/or polycarbonate (PC). PMMA is also known colloquially as acrylic glass. Both materials have a high degree of transparency (light transmission) and very good properties against environmental influences. They are weather-resistant, more impact-resistant and lighter than glass, and can also be thermoformed and bonded. Examples of applications include roofing for carports, terraces or as glazed canopies, as well as safety enclosures for production machinery, or protective screens or, currently in the case of the COVID 19 pandemic, enclosures for registration at medical facilities or checkouts in supermarkets.

material selection PMMA

Twin-wall sheets are manufactured using the extrusion process. They can be produced in virtually endless lengths in widths of over 2 m. The lengths are based on transport possibilities or the handling possibilities on the construction site. Common twin-wall sheets are manufactured in thicknesses of 4–26 mm and in delivery widths of up to 2100 mm and delivery lengths of up to 7000 mm.

manufacturing process

Twin-wall sheets have a long lifetime. Since they are usually only marginally contaminated and are also single type, the starting conditions for material recycling are very good. Specialized companies possessing the appropriate granulators and processing plants ensure a closed life cycle and valuable recycling loop.

recycling

Plastic Trash and Shopping Bags

garbage bags
vegetable bags
bread bags
plastic film
performance criteria

Garbage bags (trash bags) and vegetable bags are mostly made of plastic, while bread bags are mostly made of plastic and/or paper. The plastic garbage bag is made from a plastic film that is then tailored. It is a "throw-away" item, as it ends up in the trash after only a short time; however, this should not be the case, as a plastic garbage bag can be used several times. The use of garbage and other bags made of plastic is derived from the special properties compared to bags made of paper, because the requirement criteria for bags are elasticity, tensile strength, and moisture resistance – which clearly shows the technical advantage of plastics over paper. A drawback compared to paper is the slower biodegradability of the plastic film. However, it is not uncommon for paper bags to be coated with plastic, e.g., to improve moisture resistance, and that also interferes with their biodegradability.

material selection
PE
PP

Plastic bags are mainly made of polyethylene (PE). Garbage bags are manufactured in standard sizes from 5 l to 40 l (1.3 to 10.6 US gal) and trash bags in sizes from 70 to 250 l (18.5 to 66 US gal). Special sizes are available in all variants. Depending on the type of waste and the required carrying capacity (light household as well as office waste or heavy, wet waste from canteens), the film thickness and the plastic used vary. For light waste, a film thickness of 10 µm is sufficient; for heavy waste and large garbage bags, the thickness can be up to 100 µm.

manufacturing process
extrusion

Shopping bags and trash bags are manufactured in blown film lines – a line variant of extrusion. The extruder creates a homogeneous melt from the plastic (or several extruders in the case of multilayer films), which is then inflated to form a tube and cooled. The tube is then laid flat and is wound into a large roll on a winder by means of a take-off unit. This may be followed by further process steps, either directly or later, at a specialized company, which then lead to the tailored plastic bag (possibly with reinforced handles).

Bags made of PE and PP can be recycled easily, provided that the degree of soiling allows this to be performed economically. Garbage bags usually have only a short service life, which makes it easier to collect them by type. This is not the case for products that have a long service life. The lower the degree of contamination and the more sorted the plastics are when collected, the easier it is to recover high-quality recycled material. This recyclate can in turn be used for plastic carrier bags, and this applies in principle to all single-type films. Particularly in the case of multilayer films, middle layers are made from recyclate. In Germany, there is a system (ERDE) for eco-friendly recycling and recovery of used plastics from agriculture.

Plastic Window Profiles

Today, window frames are made of the classic materials wood, aluminum, and plastic, and combinations thereof. Window frames made of plastic have enjoyed a market share of more than 50% for decades. The plastic used must be resistant to environmental influences and ensure a long service life. In addition, the material must have a high thermal load capacity ranging from −20 °C up to more than 50 °C (−4 °F up to 122 °F) since it is used outdoors with possible direct exposure to sunlight. The required static loads (e.g., for triple glazing or large window areas) cannot be sufficiently met by the plastics in question, for which reason they are reinforced with steel profiles.

plastic window frames performance criteria

Plastic window profiles are made of polyvinyl chloride (PVC), which ideally meets the above requirements. PVC is an amorphous thermoplastic which is characterized by high resistance to environmental influences, guarantees durability and is also dimensionally stable over a wide temperature range. Due to the large permanent load and static requirements for windows, as well as the required long service life of 40 years and more, the plastic profiles must be reinforced with steel profiles. The window made of synthetic material and steel is thus a successful example of how materials complement each other. Steel alone is not suitable for window construction, because the material has much poorer thermal insulation than plastic and can also corrode.

material selection PVC

Window profiles made of plastic are produced in virtually endless lengths in the extrusion process and cut to the required further processing dimensions or the available transport capacity. The profiles are made into the finished window by reinforcing them with steel profiles, which perform an important static function, and then welding them into the finished window.

manufacturing process extrusion

Nowadays, windows made of plastic can be and are recycled on a large scale. This requires that they be released from dirt and separated from the steel reinforcements. Since the windows or the profiles are largely free of dirt and sorted according to grade, they can be easily recycled. Windows made of plastic are not replaced until they have been in use for 30–40 years or even longer. At that point, they become waste. The recycling of windows made of plastic is an established process in Germany. For example, in 2016, almost 90% of the recyclable volume of PVC windows as well as roller shutters and doors made of PVC have already been recycled.

recycling

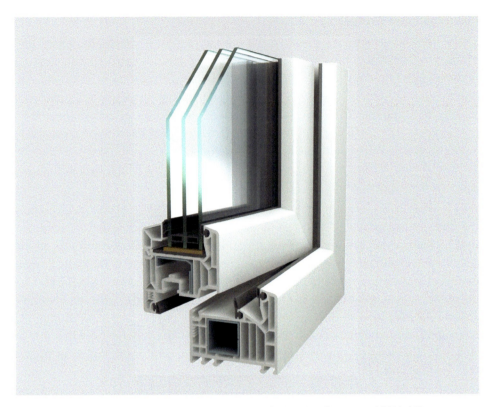

Figure 8.3 Plastic window profiles from "system SOFTLINE 82" (source: VEKA AG)

Integral Hinge and Snap Hinge

film hinge/strap hinge performance criteria

A good example of the material advantages of thermoplastics is a "film hinge", also known as a "film joint". According to their function, film hinges are "strap hinges" that do not require any additional mechanical parts for the hinge function. The film hinge (Figure 8.4) is a flexible, thin-walled hinge groove between two parts to be joined. The main requirement of the film hinge is to guarantee permanent rotation without damaging the hinge. Film hinges help in the construction of low-cost hinges. Since the component with the desired hinge function can be made from one part, assembly processes are avoided and the number of necessary parts is reduced, so saving high tooling costs.

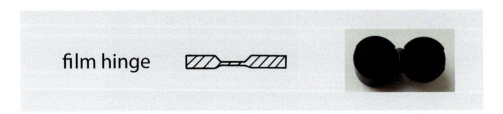

Figure 8.4 Principal sketch of a film hinge and its use as a closing cap

The film hinge is only designed for the hinge function of the two parts to be connected. If the two legs are to assume a stable closed position, then corresponding locking functions are required. Since many thermoplastics have very elastic material properties, this function is extremely easy to achieve with plastics. It can be realized during injection molding in one and the same process step.

joint function

Since a stable end position of the two plastic parts is not achieved with a film hinge (pivot point), an additional locking function is required, which can be achieved, for instance, by a snap connection (Figure 8.5).

snap-fit connection

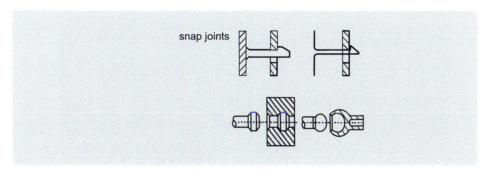

Figure 8.5 Principle sketches of snap joints

Not every thermoplastic is equally suitable at providing a permanent hinge function. As a rule, the film hinge is a more aesthetic and, most of all, more cost-effective solution than a metal solution would be. However, the load cycle resistance or the service life is often lower than that of metal. Therefore, film hinges are used only where the hinge does not have a safety function. In household articles, film hinges and snap hinges are usually made of the plastic polypropylene (PP).

material selection
PP
POM

One of the most important processes used for producing film hinges is injection molding. In addition, there are also many applications in the field of extrusion blow molding. Injection molding has a process-related advantage since the flow process gives rise to molecular orientation at the thin lip, which constitutes a flow resistance. In this particular case, this orientation is beneficial because it leads to greater strength of the material in the orientation direction, which in the case of the film hinge corresponds to the direction with the higher load. This results in a high flexural fatigue strength, i.e., an increase in the number of bending cycles and thus longer durability of the hinge. For long-term endurance, the molecule orientation should be perpendicular to the film hinge.

manufacturing process
injection molding
material orientation
extrusion blow molding

Choosing a metal solution, for example, would result in a significantly higher number of parts that would have to be assembled or mounted, which would lead to higher material and assembly costs.

hinge design
hinge calculation
lifetime

In the case of a plastic hinge, the groove that forms the *hinge* is a "weak point" in the material and must therefore be calculated precisely. The thinner the pivot point, the more precise the hinge function, because the pivot point is precisely fixed. The shorter the pivot point, the faster the resulting notch stress can lead to premature failure. Approved thicknesses for film hinges are between 0.3 and 0.8 mm for PP. The junctions to the thin section should be designed for favorable flow and provided with sufficiently large radii to prevent crack formation. Snap fits and film hinges are particularly successful examples of the possibilities of functional integration achievable with plastics. This application is also a cost-effective solution, since the product is manufactured completely in one operation during injection molding, including the snap fit and the film hinge. Downstream processes or further assembly work are not required.

■ 8.4 Performance Review – Lesson 8

No.	Question	Answer Choices
8.1	In addition to moisture resistance (e.g., in the case of wet waste), an important requirement criterion for films intended for waste bags is _____.	tear strength transparency impact strength
8.2	In addition to scratch resistance, an important requirement criterion for twin-wall sheets is the _____.	weldability transparency
8.3	Integral hinges (or film hinges) _____ be injection molded in a single operation.	can cannot
8.4	Snap connections are design elements that ensure that the parts to be connected (e.g., the two sides of a DVD box) _____.	snap together close watertight
8.5	Plastic bags compete directly with paper bags. Therefore, the _____ is/are of great importance in addition to the technical requirements criteria.	material costs transparency color design
8.6	Compared to metals, thermoplastics are _____ dependent on temporal stress and temperature.	more less
8.7	Thermosets (also called duromers) are _____ to metals such as aluminum and steel in terms of their ability to withstand stress over time and their behavior under the influence of temperature.	superior inferior
8.8	Plastic windows are made of polyvinyl chloride plastic, because PVC is very resistant to environmental influences and at the same time very _____.	temperature resistant water resistant

9 Lesson
Plastics Compounding

Subject Area	From Plastic to Product
Key Questions	Why are plastics compounded?
	What are the functions of the individual additives?
	What are the processing steps in compounding?
Contents	9.1 Fundamentals
	9.2 Metering
	9.3 Mixing
	9.4 Plasticizing
	9.5 Pelletizing
	9.6 Crushing
	9.7 Performance Review – Lesson 9
Prerequisite Knowledge	Raw Materials and Polymer Synthesis (Lesson 2)
	Physical Properties (Lesson 6)

■ 9.1 Fundamentals

Up to now, we have described how the raw material is turned into a plastic. To guarantee good processing and corresponding properties in the subsequent use of this plastic, it has to be prepared. Through compounding, the plastic acquires the necessary processing and performance properties. Figure 9.1 gives an overview of the different types of compounding.

compounding

9 Plastics Compounding

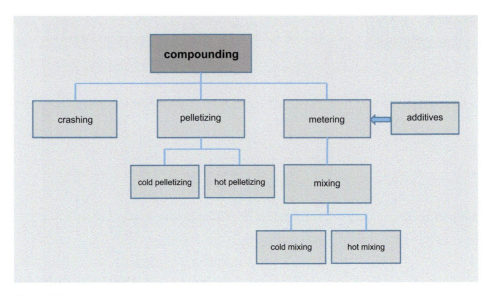

Figure 9.1 Types of compounding

functions

The compounding process performs two important functions. On the one hand, the additives, which can be present in widely varying proportions, are evenly distributed throughout the overall material, and on the other, the plastic is be converted into a form (e.g., granules or pellets) that facilitates processing.

properties

The targeted incorporation of additives into the plastic allows its properties to be changed (Table 9.1).

Table 9.1 Additives for Plastics Processing

Additives	Effect
antioxidants (heat stabilizers)	prevention of degradation reactions of the plastic through oxidation
light stabilizers	prevention of degradation reactions of the plastic through exposure to light (UV light)
lubricants	influence on the processing properties of the plastic during plasticizing
plasticizers	reduction in the modulus of elasticity
pigments	coloring of the plastic
reinforcing agents	increase in the modulus of elasticity

heat stabilizers

Let us now take a closer look at the effect of such additives, using thermostabilizers and plasticizers as examples. A heat stabilizer, for instance, enables the plastic to withstand the processing temperature without being degraded. This additive therefore facilitates the processing of the plastic.

Plasticizers make naturally hard and brittle plastics flexible and stretchable, allowing them to enter entirely new areas of application. For example, an otherwise hard and brittle plastic can be turned into a flexible, tough film. The additive thus alters the performance properties of the plastic.

plasticizers

■ 9.2 Metering

Since the precise metering (also called dosing) of the individual components is crucial when adding the additives to the raw plastic, it is necessary to measure ("meter") these components. There are two ways to meter. One is by volume ("volumetric feeding") and the other is by weight ("gravimetric feeding").

types of metering

Metering by volume suffers from the disadvantage of being relatively inaccurate, since the substances are usually present in granular form. The spaces between the grains are of different sizes, so that the actual proportion of the substance is often different for the same volumes. The advantage is the relatively low price of the metering devices.

volumetric feeding

Metering by weight, i.e., weighing, is much more accurate and can be automated much more effectively than metering by volume. Unfortunately, the required devices are significantly more expensive.

gravimetric feeding

■ 9.3 Mixing

The purpose of mixing is to distribute the additives as evenly as possible throughout the plastic without putting too much strain on it. This is usually done in discontinuously operating machines that generate a relative movement between the material particles to be mixed. A distinction is made between two mixing processes, cold and hot mixing.

mixing methods

Cold Mixing

Cold mixing is performed at room temperature, where the individual components are simply blended. An example of this blending method is the free-fall mixer (Figure 9.2), in which the mixing process is carried out solely under the influence of gravity. It is particularly suitable for mixing compounds with different particle sizes.

cold mixing

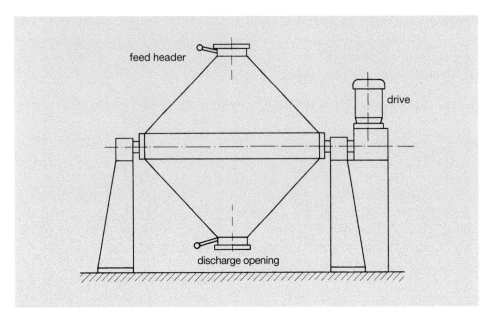

Figure 9.2 Free-fall mixer

Hot Mixing

During hot mixing, the material being mixed is heated. At temperatures of up to 140 °C (285 °F) certain additives melt and diffuse into the plastic. An example of a hot mixer is the "vortex mixer", which consists of a hot mixer and a cooling mixer (Figure 9.3).

The high-speed mixing tool of the vortex mixer generates a strong relative movement of the particles to be mixed. The resulting frictional heat and possible external heating cause the mixed material to melt. To be able to store the ready-mixed material, it is transferred from the hot mixer to the cooling mixer.

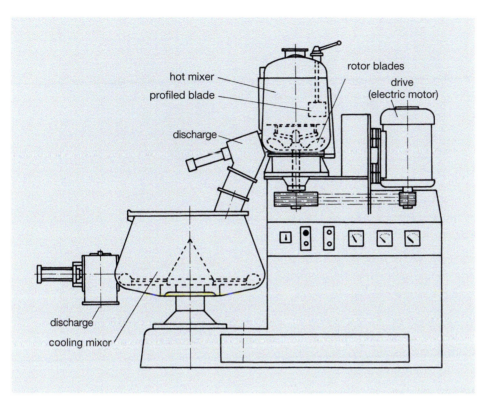

Figure 9.3 Vortex mixer

9.4 Plasticizing

Plasticizing converts the ready-mixed plastic into a suitable form for further processing. Another effect of plasticizing is further homogenization of the plastic. During this step, large quantities of additives (fillers) can also be metered in, which would be uneconomical in the hot mixer. Rolling mills, kneaders and screw machines are suitable for this operation.

homogenization
fillers

An example of continuous compounding on a shear roll mill is shown in Figure 9.4.

shear roll mill

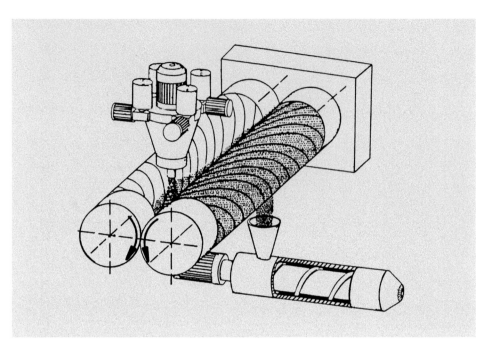

Figure 9.4 Continuous compounding on a shear roll mill

internal mixer

An example of a non-continuous or batchwise compounding unit is the internal mixer, also called a ram kneader (Figure 9.5).

It is particularly suitable for incorporating fillers, plasticizers and chemicals into rubber compounds and tough plastics. High shear and elongation forces must be applied here. A pneumatic or hydraulic ram closes the mixing chamber, which however will not be filled entirely.

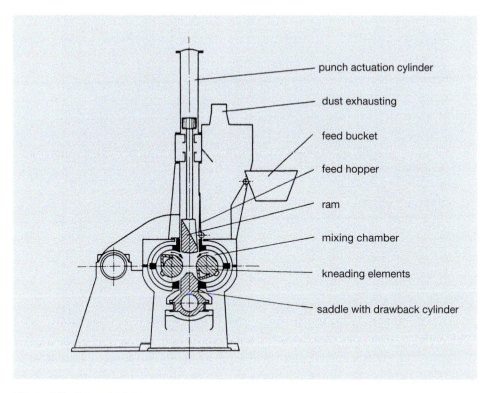

Figure 9.5 Internal mixer

■ 9.5 Pelletizing

"Pelletizing" is the process of cutting the plastic into small, free-flowing pieces. In this context, there are two variants, hot pelletizing, and cold pelletizing (see Figure 9.1).

pelletizing methods

In cold pelletizing, the plasticized plastic is first cooled and then cut into pieces (Figure 9.6). The disadvantage is that the pieces have cutting ridges and therefore do not flow as well as the hot pelletized ones, since they become wedged more easily.

cold pelletizing

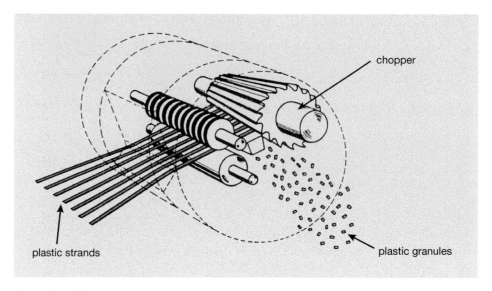

Figure 9.6 Strand pelletizer

hot pelletizing

Hot pelletizing involves plasticizing the plastic in an extruder. The extrusion die is a simple perforated plate through which the material is forced. The emerging strands are cut by a knife and the resulting pieces are cooled by air or water. The process is illustrated in Figure 9.7.

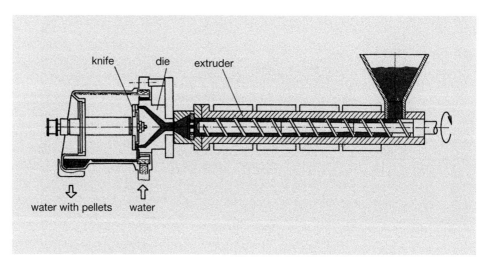

Figure 9.7 Hot pelletizing

One advantage of the process is that the still-warm particles form a burr-free shape without sharp edges, making the pellets more free-flowing.

9.6 Crushing

Crushing converts the plastic into a form that is easier to process. We have already learned about one application in the field of pelletizing.

crushing

Another increasingly important area is recycling, where rejects or collected plastic production scrap are shredded and then recycled. Cutting mills (Figure 9.8) are often used for this purpose.

plastic waste

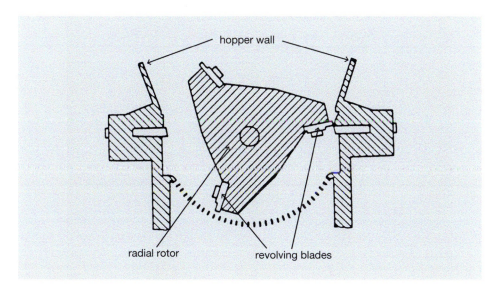

Figure 9.8 Principle behind a cutting mill

In Germany, 6.3 million tons (2019) of plastic waste are generated annually. Cutting mills are therefore important machines in the overall process of recycling plastic waste.

9.7 Performance Review – Lesson 9

No.	Question	Answer Choices
9.1	Additives are added to plastics to improve their performance characteristics and _____ properties.	inspection distribution processing
9.2	_____ are used to distribute the additives as evenly as possible throughout the plastic.	Mixers Kneaders Mills
9.3	Additives are best added according to their _____, as this method is more accurate.	weight volume
9.4	A kneader is used for the _____ of plastic.	plasticizing mixing crushing pelletizing
9.5	Hot pelletized plastic flows _____ cold pelletized plastic.	less freely than as freely as more freely than
9.6	When recycling scrap or waste plastic parts, _____ are used to crush the parts.	cutting mills strand pelletizers shear roll mills

10 Lesson Extrusion

Subject Area	From Plastic to Product
Key Questions	What characterizes the extrusion process?
	What components belong to an extrusion line?
	What specific tasks do the individual line components perform?
	Which products are manufactured by extrusion?
Contents	10.1 Fundamentals
	10.2 Extrusion
	10.3 Coextrusion
	10.4 Extrusion Blow Molding
	10.5 Blown Film Processes
	10.6 Performance Review – Lesson 10
Prerequisite Knowledge	Classification of Plastics (Lesson 3)
	Physical Properties (Lesson 6)

■ 10.1 Fundamentals

Extrusion is the continuous fabrication of an "endless semi-finished product" from plastic. The product range extends from simple semi-finished products, such as pipes, sheets, and films to complex profiles. Direct subsequent processing of the still-warm semi-finished product, e.g., by blow molding or calendering, is also possible. Because the plastic is completely melted during extrusion and obtains a completely new shape, this process is one of the primary forming processes.

continuous semi-finished product

primary processing method

■ 10.2 Extrusion Lines

A schematic illustration of an extrusion line is given in Figure 10.1.

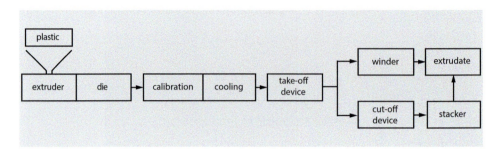

Figure 10.1 Extrusion line

The following section explains the configuration and function of the individual line components.

Extruder

homogeneous melt

The extruder is the component common to all extrusion lines and processes based on extrusion. Its task is to turn the plastic (usually pellets or powder) supplied to it into a homogeneous melt. This melt is then forced through the die under the required amount of pressure. An extruder consists of the following components shown in Figure 10.2.

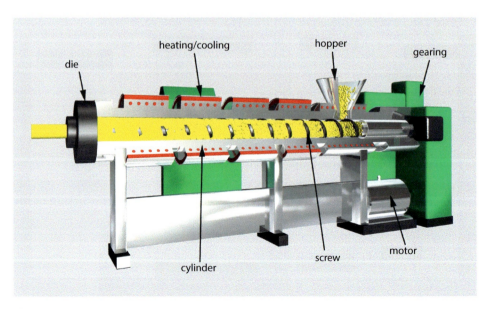

Figure 10.2 Extruder

Hopper

The feed hopper is responsible for uniformly feeding the extruder with the material to be processed. However, since the materials often do not flow freely on their own, the hopper is equipped with an additional agitator or conveyor device.

Screw

The screw performs a multitude of functions, such as feeding, conveying, melting, and homogenizing plastics, and is thus the core component of an extruder. The most common screw is the so-called "three-zone screw" (Figure 10.3) since it can be used to process most thermoplastics satisfactorily from a thermal and economic point of view. For this reason, this screw will be discussed here as a representative of all screw types.

three-zone screw

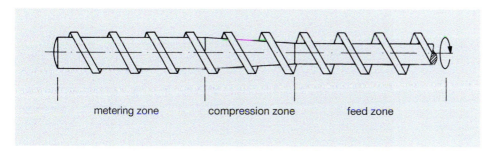

Figure 10.3 Three-zone screw

In the feed zone, the solid material is drawn in and is then conveyed forward. *feed zone*

In the compression zone, the material is compressed and melted by the decreasing flight depth of the screw. *compression zone*

In the metering zone, the melted material is homogenized and brought to the desired processing temperature. In this zone, the pressure is built up in conventional extruders. *metering zone*

A key characteristic of the screw is the ratio of the length to the outer diameter L/D. This ratio determines the efficiency of the extruder. *L/D ratio*

In addition to the commonly used three-zone screw, other screw forms are used for special applications. Regardless of their design, however, all screws, and thus all extruders, must meet the following requirements: *extruder requirements*

- constant, virtually pulsation-free conveying,
- production of a thermally and mechanically homogeneous melt, and
- processing of the material below its thermal, chemical, and mechanical degradation limits.

From an economic point of view, a high melt throughput at low specific operating costs is also required. However, these requirements can only be met if the screw and barrel are well matched, as both interact closely with each other.

Barrel

The distinction between the individual extruders is made according to the barrel (or cylinder) design (Figure 10.4).

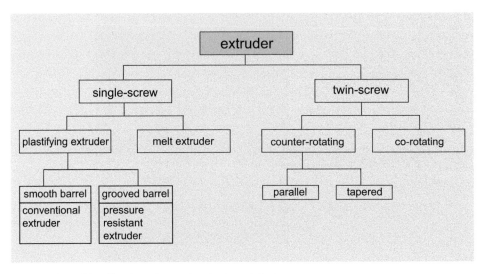

Figure 10.4 Classification of extruders by barrel design

The conventional single-screw extruder has a barrel with a smooth inner surface. It is characterized by the fact that the pressure needed to overcome the die resistance is built up in the metering zone. The feed material is conveyed by solid friction that occurs between the material particles themselves and between the particles and the barrel wall.

In the "pressure-constant" single-screw extruder, the barrel wall in the feed zone is provided with tapered longitudinal grooves ("grooved barrel"). These grooves improve the conveying and thus the compression of the material. In this case, the pressure buildup already takes place in the feed zone. However, special mixing parts must be used in the metering zone since the homogenization of the material is poorer with this type of extruder than with the conventional design.

The counter-rotating twin-screw extruder is used for powdered materials, especially for PVC. The advantage of this extruder is that additives can be mixed more easily into the plastic without subjecting the material to high mechanical or thermal stress.

In the barrel, which has the cross-section of a figure 8, the screws are arranged in such a way that closed chambers are formed between the flights, in which the material is forcibly conveyed (Figure 10.5). Only toward the end of the screw, where the pressure is built up, leakage flow occurs and the material melts by friction.

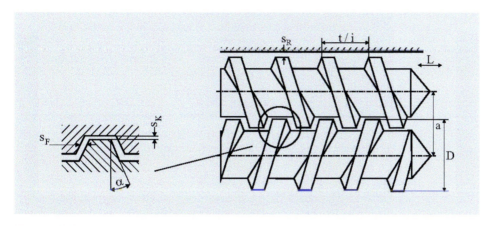

Figure 10.5 Twin-screw extruder and geometric terms

D: screw diameter, a: center-to-center distance, t: screw pitch, i: number of flights, L: screw length, s_F: back-lash, s_K: screw clearance, s_R: radial clearance, α: thread angle

The advantage of this extruder is that it allows sensitive materials to be processed at a low residence time and high temperature without exceeding the degradation limits.

The co-rotating twin-screw extruder is typically used for the compounding of polyolefins. It conveys by means of frictional coupling between screw and barrel.

twin-screw extruder, co-rotating

Heating System

The melting of the material in the extruder takes place not only by friction, but also by heat supply from the outside. It is the temperature control system that is responsible for this. The system is divided into several zones that can be heated or cooled separately. Band heaters are usually used for this purpose, but other systems are also used, such as liquid circuits. In this way, a specific temperature distribution can be achieved along the cylinder. When processing thermally sensitive materials, temperature-controlled screws are also used in some special cases.

Processed Materials

During the extrusion process, the same materials are processed that are also used for injection molding. However, there is a big difference between the two processes and, as a result, different requirements for the material. While low viscosity and high flowability are desirable in injection molding, by contrast, high viscosity is

viscosity differences

required in extrusion. This high viscosity ensures that the material stays in shape from the time it leaves the die until it enters the calibration system and that it does not flow away. In Table 10.1 lists some application examples (extrudates) that are manufactured by the extrusion process.

Table 10.1 Extrudates

Plastic	Processing Temperature Range °C (°F)	Application Examples (Extrudates)
PE	130–200 (266–392)	tubes, sheets, films, sheathings
PP	180–260 (356–500)	tubes, flat films, sheets, tapes
PVC	180–210 (356–410)	tubes, profiles, sheets
PMMA	160–190 (320–374)	tubes, profiles, sheets
PC	300–340 (572–644)	profiles, sheets, hollow bodies

Operating Principle of the Extruder

extruder principle
mixing zones

The operating principle of the extruder is similar to that of a meat grinder. As already mentioned, the material is drawn into the feed zone and conveyed further to the compression zone. There, it is compressed by the decreasing channel depth, possibly evacuated, and transformed into the molten state. In the subsequent metering zone, the material is further homogenized and its temperature is uniformly controlled (Figure 10.3). Depending on the extruder type, the pressure is built up in the feed zone or the metering zone. Since the melting process does not always provide a completely melted homogeneous melt, mixing zones (Figure 10.6) are integrated into the screw in such cases.

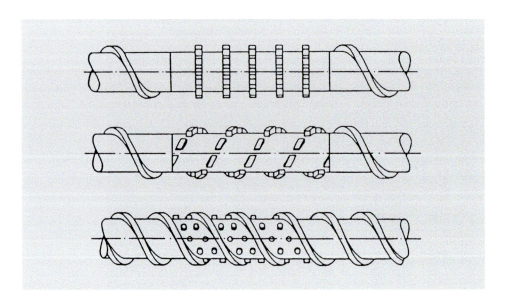

Figure 10.6 Mixing zones

Extrusion Dies

extrusion dies

While the extruder converts the material into a homogeneous melt, the die attached to the extruder determines the shape of the extruded semi-finished product, also called "extrudate". Depending on the shape, a distinction is made between different extrudates (Figure 10.7).

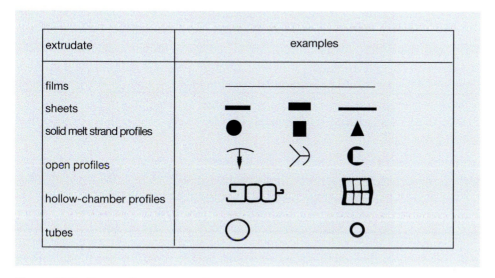

Figure 10.7 Shapes of various extrudates

Figure 10.8 shows a window corner. The window profiles are extruded and made of PVC, a material that is particularly resistant to environmental influences. That is why it is often used in the building sector.

window profile

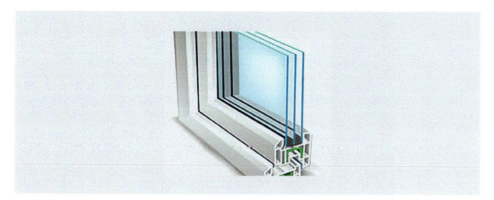

Figure 10.8 Window frame profile made of plastic (PVC) with steel reinforcement (source: KraussMaffei-Berstorff)

PVC
: PVC is a so-called engineering plastic which offers high-quality properties, such as durability and high dimensional stability over a wide temperature range (−50 to 80 °C; −58 to 176 °F), and which is also – as already mentioned – highly resistant to environmental influences. However, for static reasons (multiple glazing), the PVC profile produced by the extrusion process must be reinforced with steel profiles, otherwise the required strengths cannot be achieved. PVC window profiles have a service life of 40 years and more. They can be and are recycled to a large extent, which is discussed in detail elsewhere.

twin-wall sheets
: Twin-wall sheets for weather protection of roofing and conservatories, so-called "double-wall sheets" are made of by extrusion of a transparent plastic with high light transmission. These twin-wall sheets are made, for example, of PMMA (colloquially: acrylic glass) and PC. This is the same material that a CD is made of. PMMA and PC are amorphous thermoplastics with very good technical properties, which make them suitable as glass substitutes.

manifolds
: All molds contain a flow channel, the so-called "manifolds", through which the melt flow passes and gives the melt the desired shape. All molds are usually electrically heated. Some molds are discussed in the following.

Mandrel Support Die

Mandrel support dies (also: spider dies) are mainly used for the production of tubes, hoses, and tubular films (Figure 10.9).

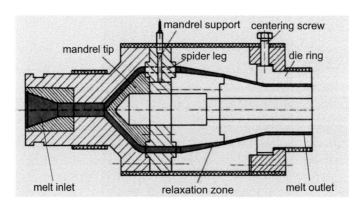

Figure 10.9 Mandrel support die

mandrel support die
: These dies have a displacer with the best-possible flow shape, which is connected to the outer wall of the flow channel via lands. It is conically shaped on the extruder side and merges into the desired inner shape of the extrudate at the die exit. Central flow to the spider die is an advantage, resulting in excellent melt distribution.

Mandrel support dies have a disadvantageous effect, as flow marks occur when the melt flows around them, which become visible in the form of local thin spots and stripes in the semi-finished product.

flow marks

To avoid such flow marks, slurring threads or spiral mandrel dies are used (Figure 10.10). The function of the slurring thread (also: wiping thread) is to superimpose a tangential component on the axial flow, thus distributing the flow marks evenly across the semi-finished product.

slurring threads
spiral mandrel dies

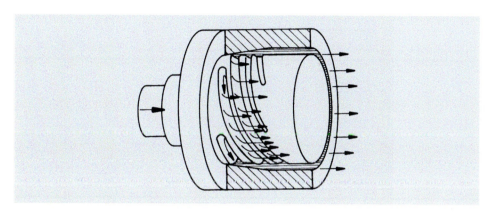

Figure 10.10 Spiral mandrel die

The spiral mandrel die has no mandrel support elements. Here, the initially radial flow is converted into an axial flow.

Flat Dies

Flat dies (also: slit distributor dies) are used to produce flat films and sheets (Figure 10.11).

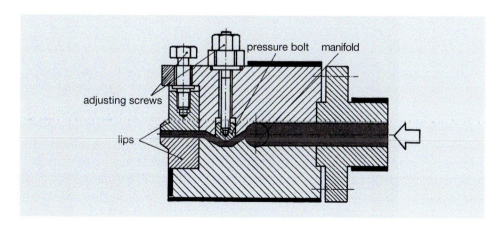

Figure 10.11 Flat die

coat hanger die

These dies first distribute the melt stream widthwise and then form it into a thin layer. In the process, the usually round melt strand first enters a distribution channel which spreads the melt out into a rectangular melt path and in most cases has the shape of a coat hanger (Figure 10.12).

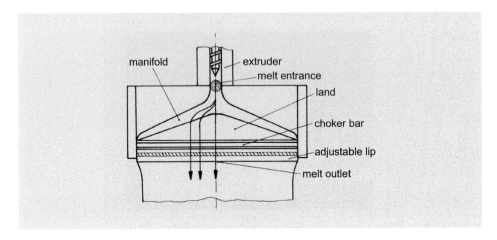

Figure 10.12 Coat hanger die

land

The melt then passes into the so-called "land" with the choker bar. The land opens into the adjustable lips from which the melt flows out of the mold.

In addition, there are numerous molds for special purposes, such as cable coatings.

Downstream Devices

calibrating device
cooling section

After leaving the extruder die, the melt must be fixed in its shape and dimensions. This task is performed by the calibrator, which operates with the aid of compressed air or vacuum. The extrudate is pressed against the calibrator walls and cools down to such an extent that it will no longer deform in the subsequent cooling section.

The length of the calibration and cooling sections must be adapted to the throughput of the extruder and to the shape of the extrudate. Sheet extrudates are cooled by rolls. By contrast, profiles, pipes, cables, and similar shapes are cooled in water baths through which the extrudate passes. Air cooling or water spray cooling are also common.

take-off device

A take-off device is connected after cooling. Its task is to pull the extrudate at a constant speed from the die through calibration and cooling. The fact that the extrudate withstands the considerable pull-off forces without deformation is thanks to the calibration and cooling section in which it was solidified on the previous sections.

The last device of an extrusion line is the cut-off and stacking device for pipes, sheets and profiles or the winding device for films, cables, threads, and flexible pipes.

cut-off device

■ 10.3 Coextrusion

The coextrusion process is used when the requirements placed on the extrudate cannot be met by one material or when material costs need to be saved. This can be achieved by combining two highly stressable (high cost) outer layers with a low-cost inner layer. The semi-finished product is then made from several layers of different materials.

multiple layers

To produce such a composite consisting different materials, each material is plasticized in a separate extruder. In a special coextrusion die (Figure 10.13), the different melts are each formed in a separate manifold. They are only brought together shortly before they exit the die and are fused together in the process. Today, it is possible to produce composites with up to eleven layers.

coextrusion die

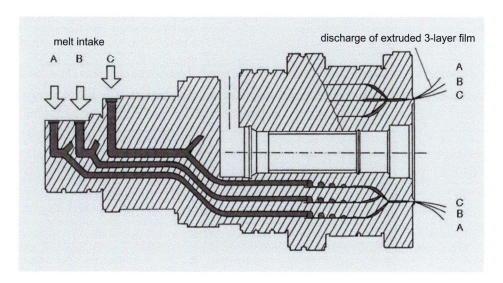

Figure 10.13 Three-layer melt manifold

Coextrusion is widely used today for multilayer cable insulation, packaging films and extrusion blow molding.

■ 10.4 Extrusion Blow Molding

products Extrusion blow molding can nowadays be used to produce hollow-bodied products from thermoplastics, such as bottles and containers for laboratory use, automotive tanks, canisters, surfboards, or heating oil tanks (Figure 10.14).

Figure 10.14 Bottles and containers for laboratory use (source: Kautex Textron GmbH & Co. KG, Bonn)

devices Two main devices are required:
- an extruder (most commonly a single-screw extruder) with a cross head,
- the blow mold and the blowing station.

process sequence The process sequence of extrusion blow molding is shown in Figure 10.15.

parison As already described, the extruder processes the plastic, resulting in a homogeneous melt. The attached cross head diverts the melt coming from the horizontal extruder to the vertical direction. Hereafter, a die forms the melt into a tubular parison. This parison now hangs freely vertically downwards.

blow mold The blow mold consists of two movable halves that has the negative shape of the part to be produced. After the parison exits the cross head, the mold closes around it and pinches it off at the bottom end. The mold is then conveyed from the machine frame to the blowing station.

10.4 Extrusion Blow Molding

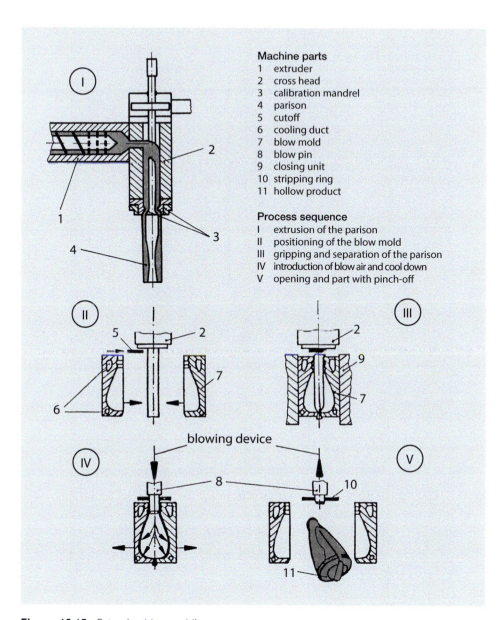

Figure 10.15 Extrusion blow molding

In the blow station, the blow pin plunges into the mold and thus into the parison. In the process, the calibration mandrel shapes and calibrates the neck area of the hollow part, while at the same time blowing air is supplied to the parison (Figure 10.16).

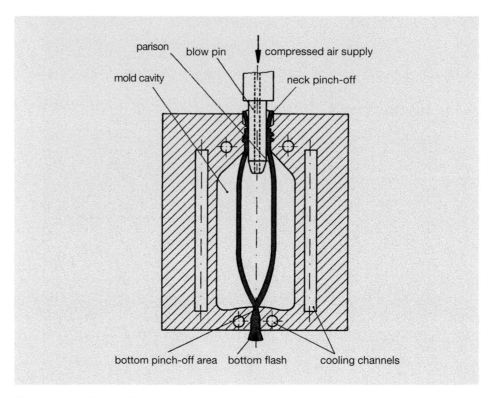

Figure 10.16 Blow mold

blow air introduction	Blow air introduction creates pressure in the parison, which causes it to inflate and press against the mold walls. This is how it acquires the desired shape. At the same time, the mold also starts to cool down.
cooling down	To shorten the cooling time, air circulation is created within the molded part by attaching an exhaust air hole to the blow mandrel. The air can thus flow out via a throttle, which serves to maintain the blowing pressure. In addition to a combination of air and CO_2, cooled nitrogen can also be used as the blowing medium.
molded part removal	After the molded part has cooled down sufficiently and thus has sufficient stability, the blow head is retracted, the mold opens, and the molded part can be removed. This part has the typical "flash" that needs to be cut off in a subsequent process.

10.5 Blown Film Process

The blown film process can be used to produce films that are mainly used in the packaging sector or in the field of construction, but also in agriculture.

blown film line

With these systems, too, the plastic is first melted and homogenized in an extruder. A die shapes the melt so that it emerges from the die gap into the shape of a circular ring. The resulting melt tube is blown up and drawn off and thus biaxially stretched. In this case, biaxially stretched means that the film tube increases in diameter because of the inflation and in length as a result of the draw-off. Stretching occurs simultaneously with cooling and ends with the solidification of the film. The cooling is done by an air blower. After this, the film tube is either wound up directly or tailored, for example cut, folded, or printed.

biaxial stretching

Most of the film production is made of thermoplastics PE (HD and LD type polyethylene) and PP (polypropylene). PE and PP films are largely used for sacks, garbage bags, bags, carrier bags, nets and fabrics, household films, stretch films, heavy shrink films, medium and fine shrink films, and adhesive tape. Other plastics used include PVC (polyvinyl chloride), PS (polystyrene) and, increasingly, PET (polyethylene terephthalate). Current blown film lines are differentiated by the film take-off, either upward (as shown in Figure 10.17) or downward. The extruder(s) for multilayer films operate horizontally in all cases.

materials
film applications
multilayer films

More and more coextruded multilayer films are being produced that can meet more complex quality requirements or ecological demands for the reuse of plastics. For example, recycled material is increasingly being used in the center layers of multilayer films. Films are produced in many different thicknesses and sizes. For example, plastic garbage bags are produced and converted from a 0.008 mm thick PE-HD film, and in the food packaging sector, 0.2 mm thick 7-layer barrier film with a 0.002 mm thick barrier layer is used.

coextrusion
recyclate

Biopolymers are also increasingly being used. These are polymers that are not based on petroleum but on cellulose, starch, lignin, chitosan and plant and animal proteins.

biopolymers

All these examples and products underline the special nature, diversity and thus the importance of the extrusion plastics processing method presented here.

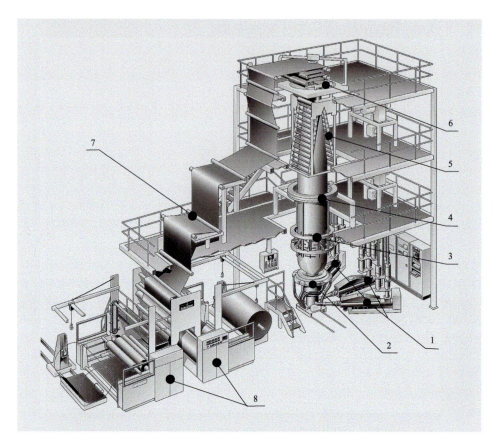

Figure 10.17 Blown film line for 3-layer film (source: Reifenhäuser).

1: extruder (3-layer), 2: blow head, 3: calibration, 4: thickness gauging, 5: flattening, 6: take-off unit, 7: web guide system, 8: winder

■ 10.6 Performance Review – Lesson 10

No.	Question	Answer Choices
10.1	During extrusion, the products are manufactured _____.	continuously / batchwise
10.2	Which component of an extrusion line is responsible for homogeneously melting the plastic? _____.	the calibrator / the extruder / the die
10.3	The most common screw design is the _____.	degassing screw / short compression screw / three-zone screw

No.	Question	Answer Choices
10.4	To avoid the extrudate "flowing away" as it leaves the mold, the processed plastic should have a _____ viscosity.	low high
10.5	The die determines the _____ of the extrudate.	length shape temperature
10.6	Coextrusion is used to produce films and sheets with _____.	a single layer multiple layers
10.7	Car tanks, surfboards and bottles are manufactured by _____.	coextrusion extrusion blow molding
10.8	In blown film extrusion, a melt bubble is inflated and taken off, thus the material is _____.	biaxially stretched being heated internally axially compressed
10.9	Solid strand profiles and open profiles are produced by _____.	extrusion blow molding extrusion or coextrusion
10.10	Tubes and hollow-chamber profiles are produced by _____.	extrusion blow molding extrusion or coextrusion

11 Lesson
Injection Molding

Subject Area	From Plastic to Product
Key Questions	What is the setup of an injection molding machine?
	What are the functions of the individual components?
	What happens during the injection molding process?
Contents	11.1 Fundamentals
	11.2 Injection Molding Machine
	11.3 The Injection Mold
	11.4 Process Flow
	11.5 Other Injection Molding Processes
	11.6 Examples and Products
	11.7 Performance Review – Lesson 11
Prerequisite Knowledge	Classification of Plastics (Lesson 3)
	Physical Properties (Lesson 6)

■ 11.1 Fundamentals

Injection molding represents the most important process for the production of molded parts made of plastic. It can be used to produce a large variety of molded parts ranging from a few milligrams to around 120 kilograms. Injection molding is one of the primary molding processes. Figure 11.1 shows a schematic diagram of the injection molding process.

injection molding

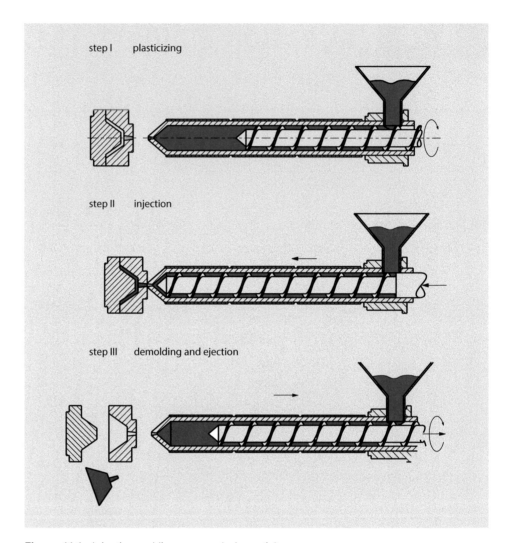

Figure 11.1 Injection molding process (schematic)

commodities
minimal finishing

Injection molding is suitable for commodities since the raw material can usually be converted into a finished part in a single operation. In contrast to metal casting and pressing of thermosets and elastomers, no flash occurs during injection molding of thermoplastics if the mold quality is good. As a result, little or no reworking is required on the injection molded part. This means that even complicated geometries can be produced in a single operation.

Plastics processed by injection molding are mainly thermoplastics, but thermosets and elastomers are also processed (Table 11.1).

Table 11.1 Plastics for Injection Molding

Thermoplastics	Thermosets	Elastomers
polystyrene (PS)	unsaturated polyester resin (UP)	nitrile butadiene rubber (NBR)
acrylonitrile butadiene styrene (ABS)		styrene butadiene rubber (SBR)
polyethylene (PE)	phenol formaldehyde resin (PF)	polyisoprene (IR)
polypropylene (PP)		
polycarbonate (PC)		
polymethyl methacrylate (PMMA)		
polyamide (PA)		

The output of parts per unit of time is decisive for economic efficiency. It is strongly dependent on the cooling time of the molded part in the mold, which in turn depends on the greatest wall thickness of the molded part. The cooling time increases with the square of the wall thickness! This must be considered for molded parts with thick walls and is of great importance to the economics. The duration between two molded parts that drop out of the machine one after the other is called the cycle time.

cooling time
cycle time

The main characteristics of injection molding are:

characteristics

- direct process from the molding compound to the finished part
- little or no post-processing of the molded part
- process can be fully automated
- high reproducibility of the molded parts
- high quality of the molded parts

■ 11.2 Injection Molding Machine

Injection molding machines are typically universal machines. The application comprises the discontinuous production of molded parts from chiefly macromolecular molding compounds, with primary molding taking place under pressure.

definition

These tasks are performed by the various components constituting an injection molding machine (Figure 11.2).

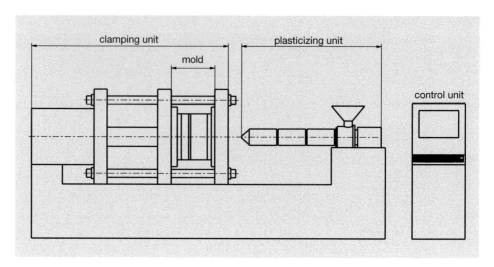

Figure 11.2 Principal layout of an injection molding machine

Injection Unit

functions

The injection unit (or: plasticizing unit) melts the plastic, homogenizes it, conveys it, meters it, and injects it into the mold. The injection unit thus has two basic tasks: firstly, to plasticize the plastic and, secondly, to inject the plastic into the mold. The use of reciprocating screw machines is common today. These injection molding machines operate with a screw that can slide back. Thus, the screw serves as an injection plunger (Figure 11.3). The screw rotates in a heated barrel, to which the material is fed from above through a hopper.

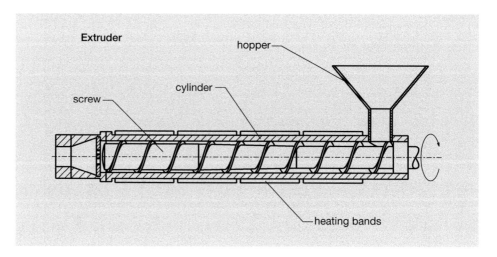

Figure 11.3 Injection unit of a screw injection molding machine

The injection unit is generally mounted movably on the machine bed. As a rule, the cylinders, screws, and nozzles can be exchanged so that they can be adapted to the material to be processed or also the shot volume.

Clamping Unit

The clamping unit of an injection molding machine is comparable to that of a horizontal press. The clamping platen on the nozzle side is fixed, while the clamping side is designed to be movable so that it can slide on four tie bars. The molds are clamped on these vertical clamping platens in such a way that the finished molded parts can fall out downwards.

The two most common drive systems for the clamping platen are: drive systems

- the hydraulically actuated toggle lever
- the purely hydraulic clamping unit

Toggle systems are used for small to medium machine sizes. The toggle lever is driven hydraulically (Figure 11.4).   toggle mechanisms

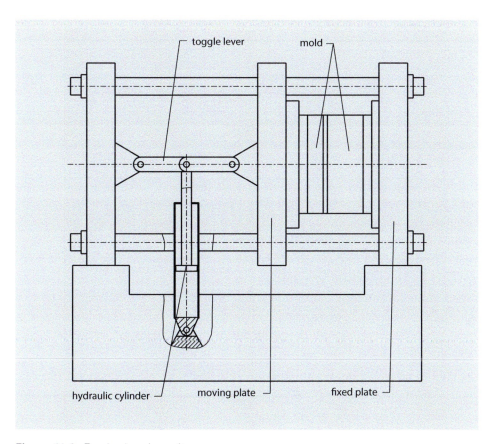

Figure 11.4 Toggle clamping unit

The advantages of this system are its self-locking operation and the fast, favorable sequence of movements and speeds. The disadvantages are possible tie bar damage or permanent deformation of the mold. This may occur if the system is poorly adjusted. In addition, the system is very demanding in terms of maintenance.

hydraulic clamping unit

There is no risk of tie bar breakage in the case of the purely hydraulic system (Figure 11.5) since the hydraulic fluid is yieldable and thus absorbs excessive deformation.

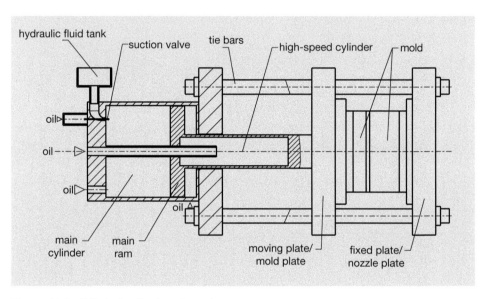

Figure 11.5 Fully hydraulic clamping unit

The advantages of this system are its greater precision, random positioning, no risk of unacceptable mold deformation and tie bar breakage. Disadvantages are its slower clamping speed, the lower stiffness of the clamping unit, due to the high compliance of the oil, and the higher energy consumption.

Machine Bed and Control Cabinet

machine bed

The machine bed is designed for mounting the plasticizing and clamping unit. It encloses the tank for the hydraulic fluid and the drive for the hydraulics. Sometimes the control and operating equipment is also incorporated directly into the machine bed.

control cabinet

The control cabinet contains the instruments, the electrical switching elements, the controllers, and the power supply system. This is the control or regulating unit of the machine. In modern machines, parameters are entered via keyboard and on-screen dialog. The microcomputer in the control cabinet controls the sequence, monitors process and production data, stores data and records the process.

11.3 The Injection Mold

The mold does not belong directly to the injection molding machine, as it must be designed individually for each part to be molded. It consists of at least two main components, each of which is mounted on one of the clamping platens of the clamping unit. The maximum mold size is given by the size of the clamping platens and the distance between two adjacent tie bars of the injection molding machine.

injection mold

The tool essentially consists of the following elements:

mold components

- platens with cavity
- gating system
- heating system
- ejection system

These elements basically meet the following requirements:

requirements

- pickup and distribution of the melt
- forming the melt into the molded part shape
- cooling of the melt (thermoplastics) or supply of the activation energy (elastomers and thermosets)
- ejection of the molded part from the mold

Figure 11.6 shows an example of an injection mold.

Molds are classified according to the following criteria:

classification criteria

- basic structure
- type of ejection system
- type of gating system
- number of cavities
- number of parting lines
- size of the mold

The costs of injection molds are very high. They generally range from €10,000 to several €100,000, which makes them only worthwhile for productions involving larger numbers of pieces.

mold costs

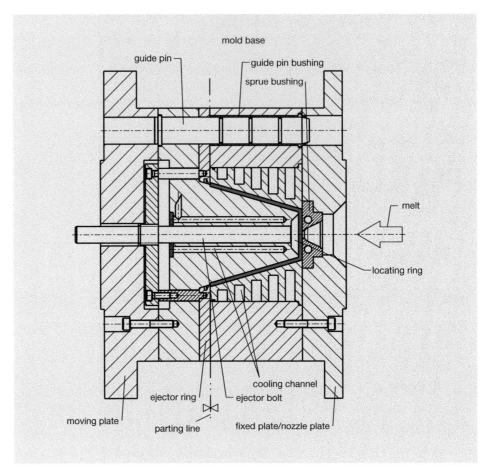

Figure 11.6 Injection mold

Ejection System

The ejector unit with the ejector plates and pins is a movable functional element. At the end of the cooling time, the mold is opened by the clamping unit. The ejector pins are moved in the direction of the molded part by a hydraulic cylinder, and the molded part is pressed out of the cavity by the ejector pins so that it falls out of the mold.

Gating

In the injection phase, the melt is forced through the gating system and fed through a gate into the mold cavity, which forms the molded part. The gating system can be designed as a heated or an unheated system.

Improved cost-effectiveness can be achieved by using several mold cavities in one mold. In this case, the gate is designed in such a way that a channel coming from the direction of the plasticizing unit branches into multiple channels leading to the individual mold cavities. The gating system should be designed in such a way that the individual mold cavities are filled as simultaneously as possible. In addition, the melt should have the same pressure and temperature when it enters the different mold cavities. This kind of gating is called "multi-gating".

■ 11.4 Process Flow

The process sequence, commonly referred to as the injection molding cycle, is shown in Figure 11.7.

injection molding cycle

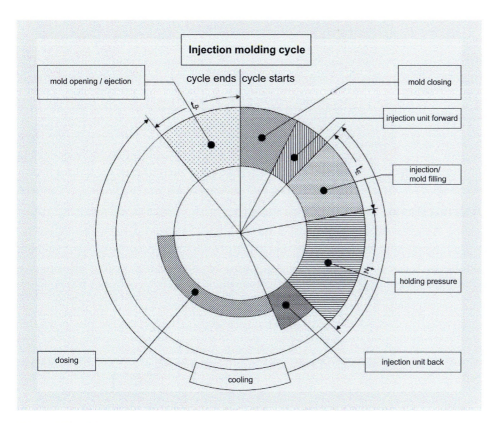

Figure 11.7 Injection molding cycle

To illustrate the chronological sequence of the individual process steps, the operations are schematically plotted against time in Figure 11.8.

chronological sequence

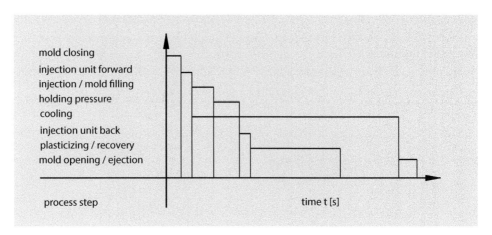

Figure 11.8 Sequence of events during an injection molding cycle

It can easily be seen here that the process steps take place one after the other, except for the important cooling process, which overlaps with other processes. The cooling process takes the longest time in the injection molding process and thus dominates the cycle time. Again, bear in mind that the cooling time increases with the square of the wall thickness!

profitability

These process steps are coordinated by the control unit of the machine and are repeated for every injection process. The cycle time should be as short as possible in order to achieve a high output rate and thus good profitability.

Metering (Feed)

screw tip

The material is conveyed by a screw, which rotates in a barrel, from the hopper towards the tip of the screw. In the process, the material is compressed and melted. While the screw is conveying material, it is simultaneously pushed back by the material that accumulates in front of the screw tip. The conveying of the material stops when the screw has reached a certain position (Figure 11.9).

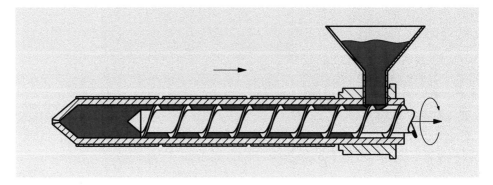

metering stroke
metering volume

Figure 11.9 Screw position after metering

At that point, enough material has accumulated in front of the screw tip to inject the molded part. The distance covered by the screw is called the metering stroke, and the volume of material in front of the screw is called the metering volume. Both parameters are reset for each mold.

Injection

During injection, the hydraulic injection cylinder drives the screw forward without rotation and pushes the metered melt through a nozzle into the mold. Due to the nonreturn valve, the screw can operate as a ram.   *nonreturn valve*
ram

The injection pressure is specified on the machine as a fixed parameter and represents an upper limit which must not be exceeded. Another parameter that must be set on the machine is the injection speed. This speed can, however, be varied during injection.   *injection pressure*
injection rate

Flow Processes within the Mold

The Flow Processes in the Mold Can be Categorized into Three Phases: *3 phases*

- Phase 1: injection
- Phase 2: compression
- Phase 3: holding pressure

During the injection phase, the mold is filled volumetrically. As soon as this is the case, the speed of the melt slows down. The compression phase begins. To compact the molded part, additional melt is conveyed into the mold. This melt is about 7%. The pressure in the mold cavity rises steeply during the compression phase. When a defined pressure level is reached in the mold, the system switches from compression to holding pressure.   *injection phase*
compression phase

During cool-down, the material in the mold cavity shrinks and new material must be added to keep the volume of the molded part constant. The holding pressure phase serves this purpose. The pressure in the molded part decreases over time, even under constant holding pressure, as the molded part solidifies more and more. Once it has dropped to atmospheric pressure, the holding pressure phase is terminated.   *shrinkage*
holding pressure phase

The time at which the changeover to holding pressure takes place is important. If the switching point is too early, the molded part will not be sufficiently compressed and sink marks will occur, while switching too late can result in overpacking of the cavity and thus flash formation occurs on the molded part.   *switching point*

After the holding pressure phase has ended, the injection unit starts directly with the new metering for the next shot.

Cooling Process

cooling time
cooling channels

The cooling time starts with the filling process and ends with demolding. This time is set in such a way that the molded part has just reached a certain temperature and is thus dimensionally stable. This process is made more efficient by cooling channels in the mold through which a cooling medium flows.

example

The CD and DVD are manufactured on a so-called horizontal injection molding machine. For a better understanding of the individual assemblies, we follow the path of the plastic through the machine. The raw material, which is usually in pellet form, is filled into a hopper into the machine and then formed into the finished product, such as a CD or DVD. This forming process is called injection molding. Figure 11.10 shows a schematic diagram of the transfer of the material from hopper to mold.

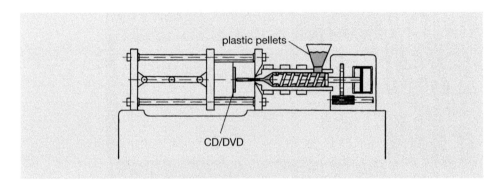

Figure 11.10 Sectional view of a horizontal injection molding machine

As already mentioned, the cooling time has the greatest influence on the production time of an injection molded part, thus also on the production time of a CD/DVD. The cooling time can be roughly estimated with a simple formula.

$$t_k \approx 2s^2 \tag{11.1}$$

t_k = cooling time s = wall thickness

The largest wall thickness of 1.14 mm is used as the wall thickness in this formula, as this determines the cycle time. A CD or DVD thus has a cooling time of approx. 2.6 seconds.

CD/DVD

After a CD or DVD has been injection molded, it is further processed. First, it is coated with a metal layer that can reflect the laser beam. It then passes through quality control and is printed. This is done in integrated systems known as production centers (Figure 11.11).

Figure 11.11 Production center for manufacturing optical data carriers (source: Singulus company)

11.5 Other Injection Molding Processes

Fluid Injection Technology

Fluid injection technology (FIT) can be assigned to multi-component injection molding in terms of the process. Instead of a second plastic melt, a process medium is used which is injected via a so-called injector into the molten core of the pre-molded part to form the cavity. This medium is gas, mostly nitrogen, in the case of gas injection technology (GIT), and water in the case of water injection technology (WIT). The primary objectives in the development of these processes were to reduce component weight, increase stability of the product as well as to reduce manufacturing costs.

cavity formation

The classic area of application for GIT and WIT is the production of long and thick-walled molded parts in which the cavity itself performs a function, such as in fluid lines or lines for other media.

GIT/WIT

Due to the significantly better cooling effect of water compared to gas, a significant reduction in cycle time of up to 70% is achieved with WIT.

cycle time

Figure 11.12 shows an example of the fluid injection process.

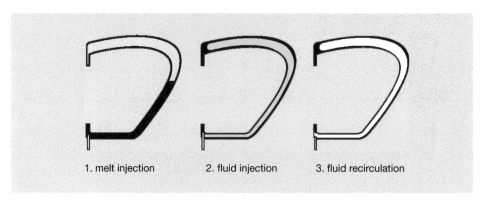

Figure 11.12 Process draft of the fluid injection technology (FIT)

After melt injection, the fluid is injected, and a chamber is created that is filled with the fluid. Once the fluid is recirculated, the molded part is ready.

Injection Molding of Thermosets and Elastomers

crosslinking reaction

Unlike thermoplastics, which solidify by cooling, elastomers and thermosets achieve their stability during processing through a chemical reaction in which crosslinking of the starting products takes place. This reaction is initiated by heat during processing in an injection molding machine.

temperatures

The injection molding process must be adapted to these vagaries, and primarily the temperatures. In the injection unit, temperatures must be kept lower than for thermoplastics to prevent premature crosslinking. The temperature range is therefore between 80 and 100 °C (176 and 212 °F) only. Since the friction generates a proportion of the heat, the plasticizing barrel may even have to be cooled in order to keep the temperatures sufficiently low.

initiation of crosslinking

It is only in the mold that heat is added to the material to rapidly advance the crosslinking. Thus, the mold is not cooled, as is the case for thermoplastics, but is heated up to temperatures of 160 to 200 °C (320 to 392 °F). This part of the cycle takes the most time. Like cooling in the case of thermoplastics, heating takes longer the greater the wall thicknesses of the molded part. It is therefore advisable to avoid the production of particularly thick parts.

parting line

With regard to the mold, special attention must be paid to ensuring that the mating surfaces, such as the parting line, are machined particularly tightly. The reason for this is the very low viscosity of the melts, which very easily leads to the formation of flash.

11.6 Examples and Products

Apart from the optical storage media already described (CD, DVD, Blu-ray disc), there are a large number of products manufactured by injection molding. In addition to extrusion, injection molding is the most important method of processing polymers. All amorphous thermoplastics (PC, PVC, etc.) and semi-crystalline thermoplastics (PE, PP, PA, etc.) can be used as materials. injection molded products

Simple applications and products made of PE and PP are, for example, wastepaper baskets, large and small household waste containers, flowerpots or planters. For products such as plant pots that are exposed to environmental influences, PP is used rather than PE. Here, the use of recycled material is also very high. Even very large waste containers, such as the gray, yellow and blue trash bins used in Germany, are produced by injection molding. They are usually made of a high-density polyethylene (HD-PE), which is resistant to many climatic effects, such as frost or solar radiation (UV rays), as the bins are often left outdoors. PE/PP applications

Higher-quality applications, such as protective shells for cell phones, which must be very thin but nevertheless dimensionally stable for weight reasons, are made of ABS, for example. With this material, dimensions and tolerances can be matched very precisely. In summer, for example, temperatures in cars can be very high, but the effect of temperature on dimensional stability is very slight for ABS material. Plastic cell phone cases are also widely made of ABS because the plastic can be printed very well. Another high-quality application is gears made of polyamide (PA) which are used in the automotive sector or model making. The best-known examples of injection molded parts made of ABS are the toy Lego bricks and Playmobil figures. ABS/PA applications automotive consumer

Telephones, printer cartridges for laser and inkjet printers, and ballpoint pens are also produced by injection molding. In cars and trucks, there are many applications, such as handles, fans and housings or covers. Very large parts, such as bumpers, are also injection molded. other applications

Another very good consumer product example is the bottle crate for beverages of all kinds. These crates are widely used and can also be recycled very well, since they are sorted by resin type and do not contain much contamination. Plastic crates for beer, mineral water, and other reusable products are now produced with a high recycled-material content. bottle crates recycling

11.7 Performance Review – Lesson 11

No.	Question	Answer Choices
11.1	Injection molding is a method of _____.	reshaping / primary processing / machining
11.2	Injection molding is suitable for the manufacturing of _____.	individual pieces / commodities
11.3	The interval between two molded parts that drop out of the injection molding machine is called the _____.	cooling time / injection time / cycle time
11.4	The injection molding process is primarily used to manufacture _____.	finished parts / semi-finished products
11.5	The _____ shapes the melt into the final part.	clamping unit / mold / plasticizing unit
11.6	The molded part _____ as it cools within the mold.	swells / shrinks
11.7	The _____ is the most time-consuming phase of the injection molding cycle.	injection time / metering time / cooling time
11.8	The _____ functions as a ram during injection into the mold.	clamping unit / toggle / screw
11.9	For injection molding of a CD with a thickness of approx. 1.2 mm, the cooling time is approx. _____ seconds.	2.9 / 5.4 / 10.8
11.10	In the case of a horizontal injection molding machine, the clamping unit and the plasticizing unit are arranged _____.	side by side / one above the other

12 Lesson
Fiber-Reinforced Composites (FRC)

Subject Area	From Plastic to Product
Key Questions	What is a fiber-reinforced composite?
	What components does an FRC consist of?
	What are the manufacturing processes for FRC components?
	What materials are used in FRC?
Contents	12.1 Fundamentals
	12.2 Materials
	12.3 Process Flow
	12.4 Hand Lay-Up
	12.5 Automated Processing Methods
	12.6 Performance Review – Lesson 12
Prerequisite Knowledge	Classification of Plastics (Lesson 3)

■ 12.1 Fundamentals

In fiber-reinforced composites (FRCs), fibers are embedded in thermoplastic or thermoset plastics. The plastic that carries the fibers is called the matrix. Suitable fibers include glass or carbon fibers, all of which have a higher modulus of elasticity than the plastic in which they are embedded. FRCs have exceptionally high specific strengths and moduli. Since the fibers increase the strength of the plastic, this group of plastics is called "fiber-reinforced". matrix
fibers

By combining plastic with fibers, the properties of both materials are combined. In particular, the strength of the plastic is increased. This combining of properties makes the fiber-reinforced composites interesting as a substitute for such materials as metals. composite

12.2 Materials

property comparison

To get an idea of the strength of fiber-reinforced composites, we will look at the modulus of elasticity of various materials in Figure 12.1.

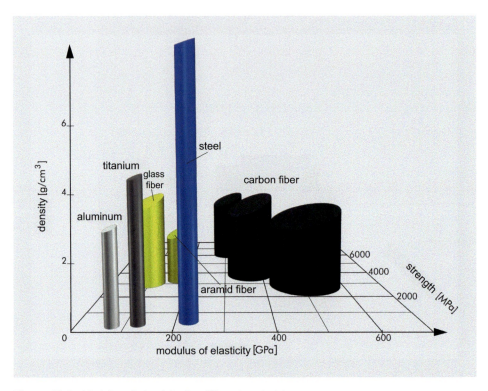

Figure 12.1 Modulus of elasticity for different materials

While the modulus of elasticity of steel is more than 200 GPa, aluminum only reaches a value of about 80 GPa. In the case of carbon fibers, the modulus of elasticity ranges from 250 to just under 600 GPa, depending on the material, and is thus 3 to 6 times higher than that of glass fibers.

isotropic

A special feature of FRC is that the increase in mechanical load-bearing capacity compared with the non-reinforced plastic occurs only in the direction of the fibers. During design and processing, attention must therefore be paid to ensuring that the fibers are positioned in the direction of the subsequent loads.

reinforcements

To meet the individual requirements for the position and effect of the fibers in the components, the fibers are optionally processed in different forms, e.g., as continuous fibers or in fabrics. The different forms are shown in Figure 12.2.

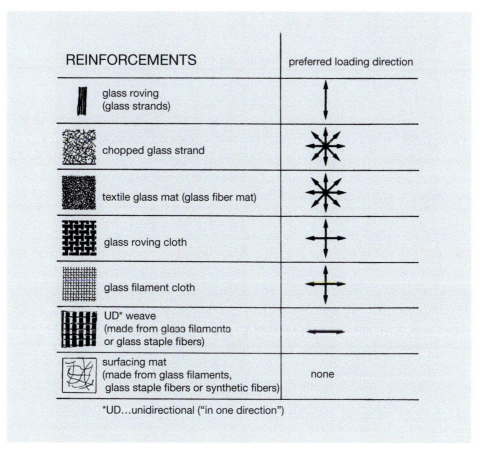

Figure 12.2 Types of reinforcement

The way in which the component is subjected to stress governs the choice of fiber reinforcement. For example, a textile glass roving can only be highly stressed in one direction, while a textile glass roving fabric can be highly stressed in two directions.

loading direction

■ 12.3 Process Flow

The production of parts made of fiber-reinforced plastics generally takes place in four steps:

- Step I: applying and aligning the fibers
- Step II: impregnating the fibers

process steps

- Step III: shaping the part
- Step IV: curing the plastic

The two steps of application/alignment and impregnation may be interchanged. The sequence of steps is different in the various manufacturing processes.

thermoset matrix

Most FRCs have a thermoset matrix. During curing of the component, this thermoset is produced through a chemical reaction in which the resin, which has previously been impregnated with the fibers, undergoes crosslinking.

hardener
promoter

To set this chemical reaction in motion at room temperature, hardeners and/or promoters are added to the resin, depending on the type. Once curing is complete, the structure of the plastic can no longer be changed, even by heating.

air bubbles

In practical use, the finished component is subjected to forces that should not damage it. However, this is only sure if the fibers adhere very tightly to the plastic. This adhesion can be affected by air bubbles on the fibers. When the plastic cures, therefore, air bubbles must no longer adhere to the fibers. If this were the case, the plastic could detach from the fiber under high stress, and the component would then be destroyed. When impregnating the fibers with resin, care must therefore be taken to ensure that no air bubbles get into the resin, or otherwise the component must be compressed and vented before or during curing.

■ 12.4 Hand Lay-Up

positive mold

The simplest way of manufacturing components from fiber-reinforced composites is "hand lay-up", a process that is widespread among craftspeople, one of the reasons being that it can be easily combined with classical model making.

In this manual process, resin and fiber mats are alternately applied to a positive mold (male mold) in layers. The fiber mat is pressed on to the mold with a laminating roller and thoroughly impregnated with resin, as shown in Figure 12.3.

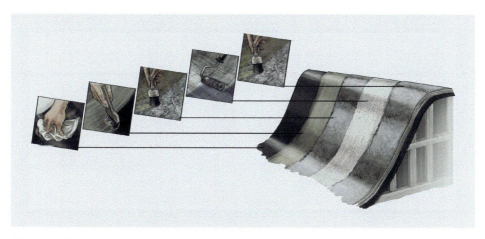

Figure 12.3 Hand lay-up

Before the actual lamination process, a release agent and a gel coat are applied to the mold. The release agent serves to improve separation of the finished component from the mold. The fine layer results in a better surface of the molded part, since the fibers do not press through this layer. — surface

Applications of the process include boat building, sailplane construction and wind turbine construction. The rotor blades of wind turbines, some of which are huge, are produced using the hand lay-up process. — application

During hand lay-up, harmful emissions are generated, which are not hazardous to health if suitable protective devices (e.g., mouth protection) are used. So, it is mandatory to use such protective devices. — occupational safety and health

After curing of the parts made of fiber-reinforced plastics, there are no longer any immediate health hazards.

■ 12.5 Automated Processing Methods

Fiber Spray-Gun Molding

In fiber spray-gun molding, the cut fibers are blown directly onto a mold by means of compressed air. At the same time, the pre-accelerated resin (I) and the resin with hardener (II) are sprayed onto the mold from two separate nozzles. The applied layer is compressed and simultaneously deaerated before the part cures. Figure 12.4 schematically illustrates the fiber spray-gun molding process. — process

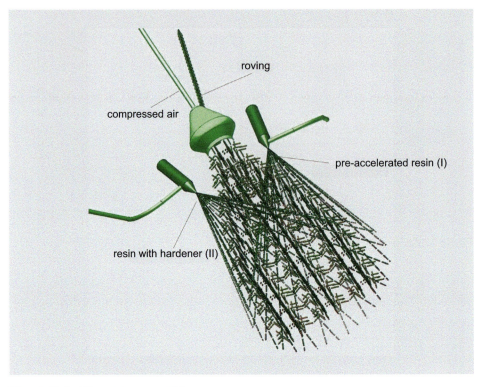

Figure 12.4 Fiber spray-gun molding

application

Since spraying produces environmentally harmful emissions such as styrene, it is advisable to use robots that work in gas-tight booths. However, the process is very often still performed manually. The fiber spray-gun process is used to manufacture bathtubs, for example.

Winding

process

In the winding process, the fiber strands ("rovings") previously impregnated with resin are wound onto a revolving mandrel. The device for guiding the fibers, the so-called "payoff eye", is moved horizontally in this process. The mandrel is thus covered with fibers in the desired manner. The method is shown in Figure 12.5.

filament direction

The guidance of the rovings and the rotational speed of the mandrel must be precisely controlled because the fibers slide off if they lie incorrectly on the mandrel. They must also have exactly the directions specified in the design to be able to absorb the forces to be encountered in the later use.

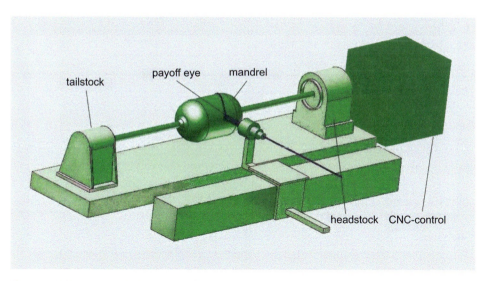

Figure 12.5 Principle behind a winding machine

In the case of more complex components, the filament guidance can be carried out point by point by hand and stored in a computer. In automatic production, the computer then controls a robot that repeats the sequence of points. — robot

Advantages of the process are its good ease of automation and reproducibility. Examples of components manufactured using this method are pipes and pressure vessels. — application

Compression Molding

The compression molding process can be used to produce large, flat parts with good mechanical properties. — process

So-called SMC or GMT compression molding compounds are processed during compression molding. SMC (sheet molding compound) consists of a resin base of chopped and/or "endless" fibers, which later hardens to form a thermoset matrix. In GMT (glass mat reinforced thermoplastics), the matrix consists of a thermoplastic material. — compression molding compounds SMC, GMT

Blanks are cut out of the SMC and GMT semi-finished products (webs) for the pressing process and stacked into packages. The package of blanks is placed in the die of the press, which is then closed and subjected to pressure. The material flows into all corners of the cavity and fills it. Figure 12.6 shows the compression molding process for SMC. — compression molding

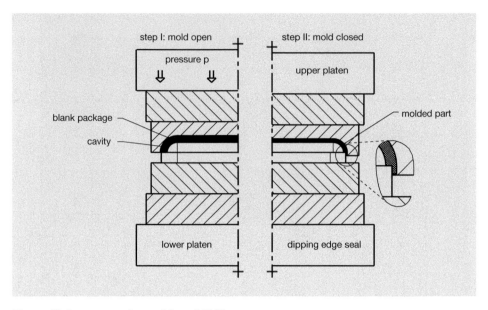

Figure 12.6 compression molding of SMC

mold

For SMC, the mold is heated, which starts the chemical reaction within the material. The chemical reaction finally cures the part. The GMT is placed in the mold at a temperature at which the plastic is still in the molten state. After being cooled in the mold, the plastic solidifies again.

part properties

Essential for the subsequent behavior of the component are the shape and position of the blank package in the mold. Both influence the flow characteristics of the plastic within the mold and thus the orientation of the fibers, which have an effect on the properties of the part, in turn.

application

Compression molding is used, for example, to produce wall elements for control cabinets or car engine hoods.

sheet molding compound procedure

The sheet molding compound (SMC) process is commonly employed in large-scale production.

It is not suitable for small quantities from an economic point of view. The investment costs of the steel or aluminum press tools are too high for small quantities and are not economically viable.

Pultrusion

pultrusion

The pultrusion or strand-drawing process is available for the continuous large-scale production of continuous filament-reinforced profiles. The pre-dried fiber rovings are soaked in a resin bath and then formed into the desired profile shape in a heated die. The applied heat causes the resin to crosslink. Figure 12.7 shows a pultrusion line: In production direction (from left to right) the process steps impregnation (dip tank), consolidation, curing, calibration (die) and take-off occur.

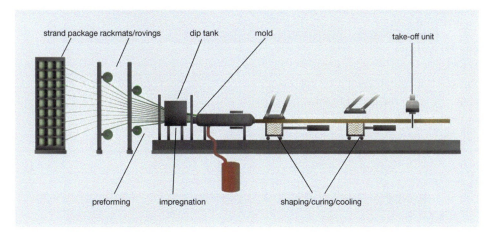

Figure 12.7 Pultrusion line

Well-known applications for pultrusion technology include sail battens, strain relief elements for optical cables and conductors and carriers for electrical engineering.

applications, products

12.6 Performance Review – Lesson 12

No.	Question	Answer choices
12.1	In the case of fiber-reinforced composites FRCs, the plastic that holds the fibers is referred to as the _____.	matrix weave mat
12.2	The modulus of elasticity of steel is 210 GPa (30,046,000 psi), while the modulus of elasticity of carbon fibers can be as high as _____ GPa.	80 GPa (11,600,000 psi) 180 GPa (261,100,000 psi) 590 GPa (855,700,000 psi)
12.3	The principal steps in the processing of FRC are: a) application and alignment of the fibers, b) _____ of the fibers, c) _____ of the part, and d) _____ of the plastic.	shaping curing impregnation

No.	Question	Answer choices
12.4	Boats made of FRC are often manufactured using the _____ process.	winding hand lay-up compression molding
12.5	Both the SMC method and the GMT method are press processes. a) The SMC process uses a_____ matrix. b) The GMT process uses a_____ matrix.	thermoplastic thermosetting thermoplastic thermosetting
12.6	The sheet molding compound (SMC) process uses blanks with a _____ matrix.	thermoplastic thermosetting
12.7	Typical components and products made of fiber-reinforced composites (FRC) are _____ and _____.	sailboats plastic windows tennis rackets plastic buckets

13 Lesson
Plastic Foams

Subject Area	From Plastic to Product
Key Questions	What are the properties of foamed polymers?
	What are plastic foams?
	How are they produced?
Contents	13.1 Fundamentals
	13.2 Properties of Foams
	13.3 Foam Production
	13.4 Examples and Products
	13.5 Performance Review – Lesson 13
Prerequisite Knowledge	Classification of Plastics (Lesson 3)

■ 13.1 Fundamentals

Plastic foams are plastics in which gas bubbles are enclosed. The space occupied by the gas bubbles in such a foam is up to 95%, while the actual plastic makes up only about 5%. Let us take as an example a cube with the volume of one dm^3. This cube of compact polystyrene weighs about 1 kilogram (2.2 lb). The same cube made of foamed polystyrene weighs only 20 grams (¾ oz).

gas bubbles volume fraction

When the gas bubbles are connected one to another, the foam is referred to as an "open cell foam" (Figure 13.1).

open cell

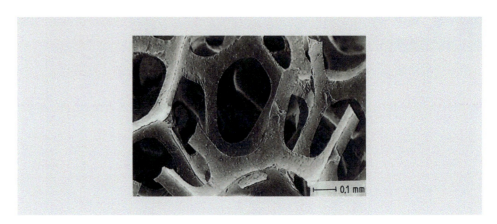

Figure 13.1 Open cell foam

closed cell

In "closed cell" foam, each gas bubble is individually present with its own "skin" (Figure 13.2).

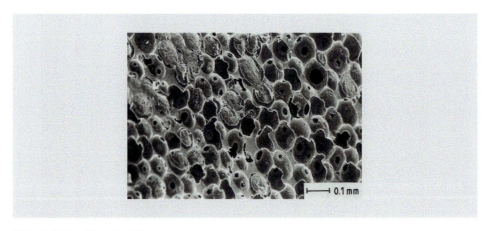

Figure 13.2 Closed cell foam

In between these two extremes, there are gradual transitions where both closed and open cells are present in the foam.

■ 13.2 Properties of Foams

cell distribution

The cell distribution can vary with the foam type. This is illustrated in Figure 13.3, which compares a polyurethane (PUR) foam with an integral (also: structural) PUR foam.

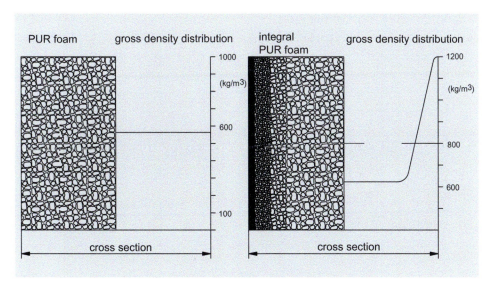

Figure 13.3 Cell distribution in PUR and in integral PUR foam

In PUR foam, the cells are evenly distributed over the cross-section. This results in an equally uniform density distribution. By contrast, "integral skin foam" has an uneven distribution of cells. While there are high numbers of cells in the center of the cross-section, their number decreases towards the edges. The outermost layer ("skin") consists practically of compact plastic. This distribution of cells is the result of a special foaming technique. The parts produced in this way have high rigidity due to the compact outer skin and are still very light. An overview of the density of various plastic foams is provided in Figure 13.4.

PUR foam
integral foam

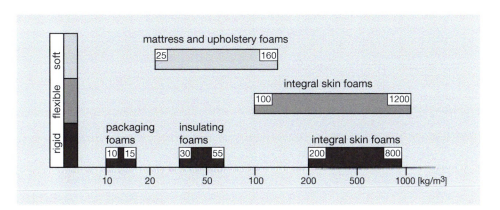

Figure 13.4 Foam densities

Plastics Suitable for Foaming

Theoretically, almost all plastics are suitable for foaming, but only a few of them are used technically. A simplified overview of plastics and processes is provided in Table 13.1.

Table 13.1 Plastic Foams

Process	Activation	Reaction type	Foaming method	Cell structure	Examples
injection molding	thermal	softening/cooling	chemical	closed cell	PVC, PE
extrusion	thermal	softening/cooling	chemical / physical	open cell	PVC, PE
multi-component system	mixture	polyaddition	chemical / physical / mechanical	open cell / closed cell	PUR
multi-component system	thermal	polyaddition	chemical	closed cell	PA, EP
two-step sintering process	thermal		physical	closed cell	PS-E (expanded polystyrene)

properties

Regardless of the plastic used and the selected process, plastic foams exhibit the following properties:

- low density
- low thermal conductivity
- favorable mechanical properties for specific weights
- simple, versatile shaping options
- easy machinability
- reduction in material

Rigidity of Foams

rigidity

One characteristic of foams is their rigidity. A simplified overview of the various plastics and their rigidity after foaming is shown in Table 13.2.

Table 13.2 Foam Rigidities

Plastic foam		Rigidity range
thermosets	polyurethane (PUR)	ductile-rigid to flexible elastic
	phenol-formaldehyde resin (PF)	brittle-rigid
thermoplastics	polyethylene (PE)	ductile-rigid to flexible elastic
	polypropylene (PP)	ductile-rigid
	polystyrene (PS)	ductile-rigid

So-called "flexible foams" (also "HR" or high-resilience foams) can be easily deformed and return to their original shape after being relieved of the load. Rigid foams can be divided into ductile-rigid and brittle-rigid foams. The ductile-rigid foams deform under load before they break.

flexible foam

If they are relieved again before fracture, some of the deformation resets. The brittle-rigid foams, on the other hand, show no deformation at all before they break.

rigid foam

13.3 Foam Production

Blowing Agents and Foaming Mechanisms

For the production of foams, blowing agents and often additives are added to the plastic. These constituents must be mixed very thoroughly in order that imperfections and irregularities in the foam may be avoided. As foaming begins, the mixture must be able to flow freely. Once the bubbles formed by the blowing agent have attained the desired size, they must be fixed, which is done by hardening the plastic.

At the beginning of the foaming process for thermosets, the plastic exists as a resin with very little to no cross-linking and low inherent viscosity. Fixing of the bubbles as they form is done by the reaction and resulting cross-linking of the plastic. The viscosity of the plastic increases rapidly during this reaction. Thermoplastics, by contrast, must be melted in order to be foamed and the bubbles become fixed by the solidification of the plastic when it cools down.

thermosets
thermoplastics

The blowing mechanisms by which the bubbles are formed can be divided up as follows:

foam generation

- mechanical foaming
- physical foaming
- chemical foaming

mechanical	In mechanical foaming process, the bubbles are formed either by introducing a gas by means of a stirrer or by injecting the gas into the plastic melt under high pressure.
physical	In physical blowing processes, a low-boiling liquid evaporates due to heat and forms the bubbles. The most widely used physical blowing agent is nitrogen.
chemical	In chemical blowing processes, the blowing agent reacts under the effect of heat-releasing gases that form bubbles.

Technical Versions

Two different processes are used to mix the components for foaming:

- low-pressure process
- high-pressure process

low-pressure mixer	One option is the low-pressure mixer, where mixing is performed by mechanical stirring. The advantage of this is that pressure is only necessary to convey the constituents through the lines. A disadvantage of the process is that only a relatively small quantity can be mixed and conveyed per unit of time. The process is therefore unsuitable for plastics that react very quickly. Another is that the mixture only flows out of the mixing chamber under its own weight. Thus, foaming can only be done with molds into which the material can be poured without additional pressure.
high-pressure mixer	The other option for mixing is the high-pressure mixer. In this process, the components of the plastic collide with each other under high pressure in the mixing chamber and are thus vortexed. The advantage of this process is that even fast-reacting plastics can be swirled, since the throughput per unit of time is very high. The mixture does not begin to react until it has been rapidly introduced into the mold. Due to the pressure, it is also possible to use closed molds into which the mixture must be injected. Figure 13.5 shows a comparison of the two mixing processes.
pressure	A comparison of the two processes shows that the high-pressure process, with a pressure buildup of up to 300 bar, requires significantly more technical effort than the low-pressure process.
molds	The molds for the parts made of plastic foams are of various types. For semi-finished products from which cushions or insulation are produced, a continuously running paper tub is used which is open at the top. It is shown in Figure 13.6.

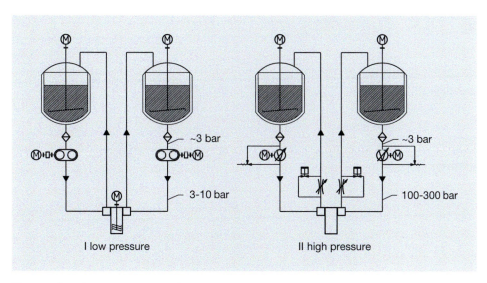

Figure 13.5 Low-pressure and high-pressure mixers (1 bar ≈ 1 atm)

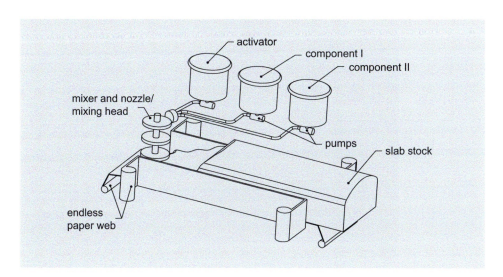

Figure 13.6 PU slab production

Another alternative is a mold similar to an injection mold. The mixture is injected into this mold by means of a high-pressure mixer until one-third of the mold is filled. Then the mixture begins to foam and completely fills the mold. This process is known as "RIM". RIM stands for "reaction injection molding" and indicates that the molded part is formed by a combination of injection and reaction. This process is used, for example, to produce automotive dashboards.

RIM applications

■ 13.4 Examples and Products

Products as diverse as mattresses in household articles, car seats, automotive dashboards or dashboard consoles, insulating boards, insulating materials, and sealing materials in the construction sector, but also yoga mats in the sports sector and soles of sports shoes, are made of plastic foams.

Figure 13.7 shows an application of a plastic-foam component from the automotive industry.

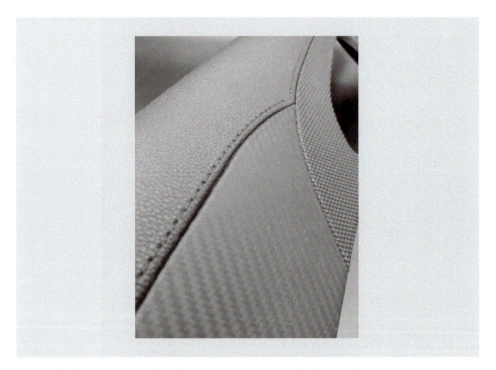

Figure 13.7 Interior door trim (foamed component with film lamination)

Plastic foams are often coated (e.g., laminated) with other materials, e.g., for a better feel. Therefore, they are often not easy to recognize.

13.5 Performance Review – Lesson 13

No.	Question	Answer choices
13.1	In plastic foams, _____ are entrapped within the plastic.	fillers gas bubbles
13.2	Plastic foams are _____ compact plastics.	lighter than heavier than exactly as heavy as
13.3	The gas bubbles of PUR foams are distributed _____ throughout the plastic.	uniformly not uniformly
13.4	There is a considerably _____ number of cells in the center of integral skin foams than at the edges.	lower higher
13.5	The rigidity is _____ for all plastic foams.	the same not the same
13.6	In the production of plastic foams, a distinction is made between mechanical, physical, and chemical _____.	mixtures foaming methods
13.7	The _____ can also be used to process fast-reacting plastics.	low pressure process high pressure process
13.8	The RIM process is similar to _____.	extrusion injection molding

14 Lesson
Thermoforming

Subject Area	From Plastic to Product
Key Questions	What steps are involved in the thermoforming process?
	What plastics can be thermoformed?
	What are the different processes?
Contents	14.1 Fundamentals
	14.2 Process Steps
	14.3 Technical Installations
	14.4 Examples and Products
	14.5 Performance Review – Lesson 14
Prerequisite Knowledge	Classification of Plastics (Lesson 3)
	Deformation Behavior of Plastics (Lesson 4)

■ 14.1 Fundamentals

Thermoforming is the forming of plastics under the action of heat and pressure or vacuum. There are various techniques for this process. The use of compressed air and/or vacuum to apply force has become the most common method of forming thermoplastics.

thermoforming

The general process flow is as follows. The plastic is heated to a temperature at which it becomes thermo-elastic or rubber-elastic (Figure 14.1). It is then reshaped and cooled down.

process flow

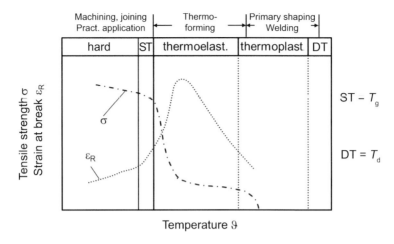

Figure 14.1 Phase diagram of amorphous thermoplastics

thermoplastics

Since thermoplastics can be converted from the solid to the thermo-elastic range by heating, these are the only plastics suitable for this type of processing. In contrast, thermosets, for example, which do not become thermo-elastic again when heated, cannot be processed in this way.

semifinished products

The films and sheets that are mainly processed have a thickness of between 0.1 and 12 mm. The material, also called "semifinished product", is available either as a single sheet or as rolled sheet stock.

PP, PS, PET, PLA

Almost all thermoplastics can be thermoformed. Since thermoformed products (such as beverage cups) are used in large quantities in the packaging sector, the cost issue plays a dominant role. For this reason, the low-cost commodity plastics PP and PS but also PET are often chosen. Their material properties suit most applications very well.

In the packaging sector, however, there is also a noticeable trend toward sustainable recyclable packaging. Here, for example, solutions consisting of plastic-cardboard combinations are used to produce cups, trays and blisters. Bio-based materials such as PLA (polylactide) are also employed. These are made up of many chemically bonded lactic acid molecules.

14.2 Process Steps

The process consists of three steps:
- heating
- forming
- cooling

In the first step, the semi-finished product is heated up. For this purpose, three possible heating methods are available: heating by convection, by contact or by infrared radiation.

Infrared radiation heating is the most widely used method, as its energy penetrates directly into the internal regions of the plastic. Consequently, the material heats up very quickly and evenly, without the surface being damaged by overheating.

The second step is the forming of the part, where the plastic is stretched. The heated semi-finished product is clamped in a support and pressed or pulled into or onto a mold by means of compressed air or vacuum. One disadvantage of the process is that only the side of the molded part that is in contact with the mold acquires the exact shape.

For this reason, a distinction is made between "positive" and "negative" processes, depending on whether the inner or outer side of the part is formed with precision. The negative process is shown in Figure 14.2.

In the negative process, the semi-finished product is *drawn into the mold*. In the positive process, the semi-finished product is *drawn onto the mold*. During this process, the semi-finished product is held in place and becomes stretched. This results in uneven wall thicknesses for the components, with particularly thin corners.

negative process

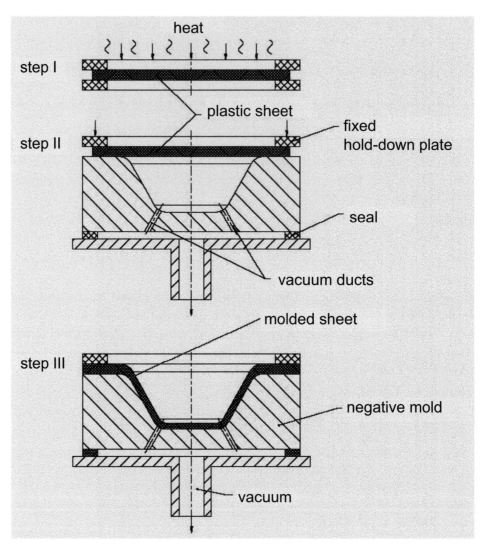

Figure 14.2 Negative vacuum thermoforming process

prestretching

In order to minimize this effect, the semi-finished product is often prestretched before the actual molding process. In the negative process, this prestretching is done by an assisting plug. In the positive process, it is done by "inflating" the semi-finished product. As an example, the positive process with prestretching is shown in Figure 14.3.

cooling

The third step is cooling. It starts as soon as the heated semi-finished product touches the tool, which has a lower temperature. To accelerate the cooling time, the tool can be specially cooled, e.g., for series production. This cooling can be accomplished with a blower, for example.

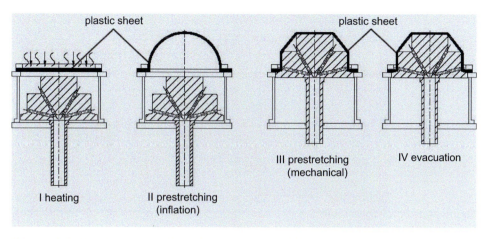

Figure 14.3 Positive vacuum thermoforming process with prestretching

14.3 Technical Installations

The technical realization of the individual process steps takes place either in single-station or multi-station machines. In a single-station machine, the technical devices are in motion, whereas the semi-finished product remains in the same position from heating through to removal from the mold (Figure 14.4).

single-station machine

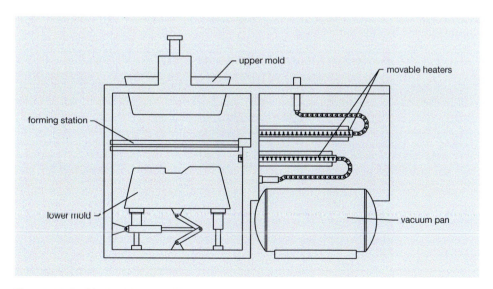

Figure 14.4 Single-station machine

multi-station machine

In the multi-station machine, the semi-finished product moves continuously from one technical station to the next (Figure 14.5).

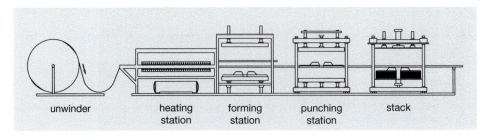

Figure 14.5 Multi-station machine

The disadvantage of the single-station machine is its long cycle time, which is the sum of the various times for the individual steps, while the cycle time for the multi-station machine is equal to the time required for the longest work step.

■ 14.4 Examples and Products

applications

These thermoforming processes are widely used to produce packaging, such as yogurt cups, but even for large parts such as swimming pools or automotive parts.

Figure 14.6 shows some products from the packaging sector that are typically manufactured by thermoforming.

Figure 14.6 Thermoformed products: Drinking cups made of PLA, PET, PP and PS, various coffee capsules and packaging made of a plastic-cardboard combination (source: ILLIG Maschinenbau GmbH & Co. KG, Heilbronn)

waste
recycled material

Yogurt cups or drinking cups, for example, are very often disposable products that only have a short lifetime and are mass-produced. Therefore, it is important that these products do not end up in a landfill but are instead recycled. In the case of drinking cups, for example, which were filled with water before they ended up in the trash, recycling is easy because they don't have a large amount of dirt. In the case of yogurt cups, on the other hand, where the dirt content (residual yogurt) makes up the largest proportion of the recyclable material by weight, it is more difficult to produce a cost-effective recycled material because expensive cleaning processes – as additional production steps – must be added.

As mentioned at the beginning of the lesson, the trend in packaging is clearly toward solutions such as plastic-cardboard combinations for cups, trays and blisters, or the use of bio-based materials such as PLA.

14.5 Performance Review – Lesson 14

No.	Question	Answer Choices
14.1	In thermoforming, the plastic must first be _____ before it can be reshaped.	cooled down heated up melted
14.2	Only _____ can be thermoformed, as they are the only type of plastic that becomes rubber-elastic when heated.	thermoplastics elastomers thermosets
14.3	The most common heating method used in the thermoforming process is _____.	convection contact heating infrared radiation
14.4	During thermoforming _____ of the part is/are precisely formed.	only one side both sides
14.5	To avoid uneven wall thicknesses, the semifinished product is _____ before the forming process.	preloaded prestretched clamped
14.6	The cycle time for multi-station machines is _____ the cycle time for single-station machines.	shorter than longer than equal to
14.7	Plastic drinking cups are typically made of the thermoplastics PP and _____.	PVC PS

Lesson 15
Additive Manufacturing

Subject Area	From Plastic to Product
Key Questions	What is additive manufacturing?
	Which materials are used?
	What processes are used in additive manufacturing?
Contents	15.1 Fundamentals
	15.2 Plastics for Additive Manufacturing
	15.3 Operating Steps and Process Parameters
	15.4 Examples and Products
	15.5 Performance Review – Lesson 15
Prerequisite Knowledge	Plastics Fundamentals (Lesson 1)
	Classification of Plastics (Lesson 3)
	Deformation Behavior of Plastics (Lesson 4)

■ 15.1 Fundamentals

Additive manufacturing (AM), which is also referred to as *layer manufacturing*, is a manufacturing process in which a part/component or a semi-finished product is built up layer by layer without the use of a molding tool. Additive manufacturing thus differs from other manufacturing processes that produce parts/components by means of machining or primary forming processes such as casting, compression molding, injection molding, extrusion or forming. In common usage, additive manufacturing is also referred to as *3D printing*. However, 3D printing represents a special variant of additive manufacturing. Here, a liquid binder is selectively printed onto a powder material by means of a nozzle head, which bonds the powder at this point, whereby a model is built up layer by layer.

additive manufacturing (AM):
3D printing
layer manufacturing
rapid manufacturing (RM)

rapid technologies

Since additive manufacturing makes it possible to build complex geometries in a very short time, it is also a "rapid technology". The word "rapid" refers to the speed of the processes, which have been used since the 1980s to produce prototypes and now also tools in rapid tooling (RT) and workpieces/products in rapid manufacturing (RM). However, the speed results not so much from the manufacturing process itself – 3D printing is significantly slower than an injection molding cycle. Rather, it results from the elimination of tooling production, which generally takes several weeks. For this reason, additive manufacturing processes are particularly suitable for the rapid production of individual pieces and small series, where the production of a molding tool is too time-consuming and too expensive. Industrial 3D printers are already producing components in small batches, for example for the automotive industry.

The basic workflow of additive manufacturing is shown in Figure 15.1 using the example of a hook.

Figure 15.1 Workflow and process steps of additive manufacturing (AM) (source: IKV)

3D model
slices
manufacturing

The basis for additively manufactured items is a 3D CAD model generated on the computer. Here, production-related changes (e.g., shrinkage allowance, support structures) must be considered. Then so-called STL files are generated, which represent the surface of the CAD model in the form of triangles, and "slicing" (the cutting into thin layers) is carried out. This step also includes path line planning, which determines the location and sequence of the elements to be applied. Depending on the selected manufacturing process, further parameters, such as the nozzle temperature or the layer height, may have to be defined.

data transfer
surface structure

The process parameters having been selected, the resulting layer model is sent as machine code (G code) to the AM system, which applies the individual layers on top of each other. In this way, complex geometries with undercuts and cavities can be generated in one piece and without additional working steps. So, once a CAD model is generated, products can be manufactured worldwide (using the same manufacturing technology from the same manufacturer) simply by transferring data. Consistent, reproducible quality is currently an open challenge to be addressed. Depending on the accuracy of the selected manufacturing process, however, the finished objects have a stair-stepping surface, which makes "post-processing" necessary to an extent depending on the requirements profile.

If a component is to be manufactured with large overhangs or undercuts, for example, it is necessary to create support structures. These usually must be removed during more or less complex post-processing. However, there are also systems that allow support structures made of a different material. For example, there are support materials that can be dissolved in certain solvents. This makes it easier to remove the support structures and thus improves the cost structure. The production of support structures extends the manufacturing time.

supports

Additive manufacturing of today has a wide range of materials at its disposal, and the various processes differ in their suitability for processing certain materials. In addition to various plastics, composites and ceramic materials, the production of objects made of metals is also possible using the various additive manufacturing processes. Thus, additively manufactured components can also be employed in areas that impose special demands on their mechanical and thermal properties, such as heat-resistant components for use in aviation. In addition, the use of biological materials also offers new possibilities for generative processes, for example in the field of medical technology. Special machines are also capable of generating a component consisting of different materials and thus combining their properties.

material diversity

The possible applications and fields of use for additive manufacturing processes are diverse. The processes can be classified into the following four types:

- rapid prototyping
- rapid tooling
- rapid repair
- direct manufacturing

Rapid prototyping refers to the production of functional prototypes whose component properties do not correspond to those of the final product.

rapid prototyping

Rapid tooling refers to the manufacturing of tools using the methods of 3D printing.

rapid tooling

Rapid repair refers to the repair of wear parts by applying new material layer by layer.

rapid repair

Direct manufacturing means the production of final components that can be installed directly or after final machining.

direct manufacturing

There is no doubt that additive manufacturing processes have revolutionized our industrial and craft manufacturing capabilities. Once the three-dimensional components have been manufactured on the basis of digital geometry and process

saving resources

data, they can be produced rapidly all over the world on the appropriate machines without delay and on site. Long and expensive transport routes are therefore eliminated, which also reduces transport-related CO_2 emissions. Additive manufacturing technologies can thus contribute to reducing the consumption of resources.

advantages of AM

Additive manufacturing processes thus ensure the direct conversion of 3D CAD data (of the virtual component) into a real (physical) component and have the following advantages:

- additive manufacturing processes enable a very high degree of geometrical freedom;
- the layer geometry is generated from the 3D CAD data;
- no product-specific tool is required;
- the cost-intensive setup processes required with other manufacturing processes are omitted;
- material attributes are derived during the build process;
- manufacturing processes can be automated to a large extent.

disadvantages of AM

However, additive manufacturing is not a technology that could replace all other manufacturing processes; it also has certain weaknesses. First, there are the necessary *post-processing steps* if, for example, a high surface quality or compliance with tolerances is required; second, long process times can arise because the component is generated layer by layer. It is the geometry or height of the component that determines the production time.

Additively manufactured components have *strongly anisotropic mechanical properties* due to the layer structure. This should be considered when designing the component, because the arrangement of the component in the production line has an effect on the production time and thus on the cost structure of the component. In addition, a significantly smaller range of materials is available for additive manufacturing than for injection molding, the *manufacturing accuracy is by far not as precise*, and the process fluctuations are significantly higher.

■ 15.2 Plastics for Additive Manufacturing

AM, as applied to plastics processing, is divided into the categories of *polymerizing processes*, *plasticizing* processes, and *powder-bed-based* processes. Thermoplastics are used in the latter two processes.

fusibility

The focus of this lesson is on plastics. The ease with which additives can be processed depends to a large extent on their response to heat. Since fusibility without thermal decomposition is an important requirement, the polymers processed as selected from the groups of thermoplastics and thermoplastic elastomers.

Since the different polymers also have varying chemical and physical properties, such as different melting temperatures, the right additive process should be selected for processing to achieve the best possible result. To create a compact object, it is essential not only to place individual layers on top of each other, but also to bond them together. The 3D object is created on a build platform which, in a number of processes, is lowered by a specified layer thickness after each process step, i.e., after the generation of every single layer, so that the next layer can be applied. The exact reproducibility of the relative movement between the plasticizing unit (in the case of the plasticizing processes) and the build platform is therefore of decisive importance for the quality of the part.

process
single layers
platform
layer thickness

Almost all thermoplastics can be used for additive manufacturing (Figure 15.2).

plastics for AM

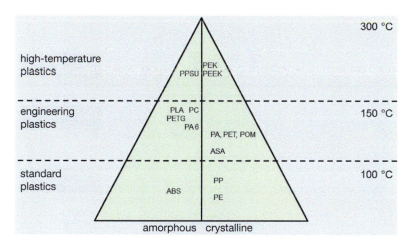

Figure 15.2 Operating temperature range and classification of plastics for additive manufacturing

Plasticizing processes can be applied to a large selection of plastic types. In addition to widely used plastics such as acrylonitrile-butadiene-styrene copolymer (ABS), polyamide (PA), polyethylene terephthalate (PET) and polypropylene (PP), special high-performance plastics such as polyether ether ketone (PEEK) and fiber-reinforced composites may also be used.

■ 15.3 Operating Steps and Process Parameters

AM processes

A large number of additive technologies are now in existence. Not all of them are used for plastics processing. Figure 15.3 summarizes the main processes according to the type of material, the operating principle and the materials used.

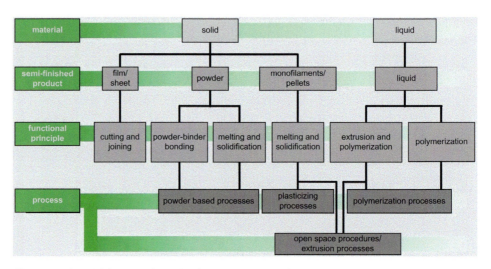

Figure 15.3 Additive manufacturing in plastics processing

We shall briefly discuss stereolithography (SLA), which is one of the pioneering AM processes, the powder process (SLS) and the plasticizing process (FLM), which is becoming more and more important, for our focus in the plastics industry.

stereolithography (SLA)

In stereolithography (SLA), liquid resins are cured by a laser beam due to their photosensitivity. The build platform is completely covered with liquid resin and the laser beam travels along the path specified in the CAM model. Where the laser beam encounters the liquid resin, the latter solidifies through polymerization and at the same time initiates bonding to the preceding layer. After each layer has cured, the platform sinks by the layer thickness, a wiper moves over the resin bath to flatten and recoat the surface. The next layer can then be polymerized by the laser. To finish the components, downstream steps are required after the actual printing process (e.g., complete curing under UV light). The surface of the components produced by the SLA process is very good and reproduces fine details.

selective laser sintering (SLS)

In selective laser sintering (SLS), energy is applied to a plastic powder by a laser, causing it to melt in a locally defined manner. The powder particles weld with neighboring particles and solidify after cooling. When a layer has solidified completely, the build platform is lowered by the next layer thickness. A new powder layer is applied (recoating) with the aid of a roller that spreads the powder uniformly and the process starts again. Care must be taken to ensure that, when energy is applied by the laser, the layer underneath bonds with the new layer by melting slightly. This results in a welded joint of individual layers, which thus form a compact structure. Various polyamides, in particular PA12, PS and PEEK, lend themselves to this process.

Plasticizing processes are of increasing importance within AM, as they are widely used in the consumer sector. They are sometimes also called extrusion processes, although no extruder is used in the actual sense. In the plasticizing unit, the plastic is melted by heat conduction (and not by shear as in extrusion) and forced through a die. This forms a strand which is deposited on a build platform according to the defined geometry and solidifies there. These processes are known, for example, as "fused layer modeling" or "fused layer manufacturing" (FLM). The first commercialized FLM process was that of "fused deposition modeling (FDM)" (Figure 15.4).

fused layer modeling (FLM)

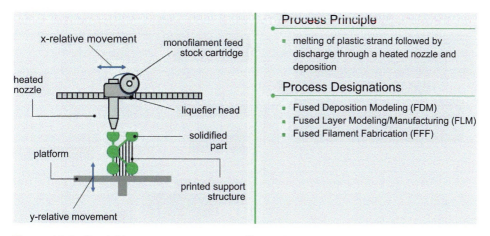

Figure 15.4 Plasticizing processes, as exemplified by fused deposition modeling (FDM)

The raw material for plasticizing processes is available as a plastic filament with a diameter of, e.g., 1.75 mm (monofilament) wound on a spool. During the manufacturing process, the filament is passed through the heated plasticizing unit, where the plastic softens due to heat conduction. The plasticizing unit, which moves horizontally in the x and y directions, has one or more nozzles. The plastic melt is applied to the build platform or substrate through these nozzles. Strands applied one on top of the other solidify to form a firm bond as soon as the material cools to below the solidification temperature.

filament

surface roughness	Items manufactured in this way usually have a clearly layered and therefore significant stair-step structure. Their surfaces are quite rough, depending on the selected process parameters, more especially the set layer height as well as the nozzle bore diameter. The specification profile may therefore make it necessary to rework the finished part.
part characteristics anisotropy strength	The properties of a part play an essential role in technical products. This imposes high demands on the production equipment and the operating temperature ranges. Adequate bonding of the individual strands and of the layers with each other is a prerequisite for the strength of the components. Due to the layering process and the associated growth of the layers in a certain direction, the mechanical properties of the components are strongly direction-dependent. Usually, a component exhibits significantly lower strength in the build-up direction (z-direction) than in the x-y direction (this is referred to as "anisotropy"). The subsequent intended use of a component and the associated mechanical and thermal loads must therefore be considered when selecting the build-up direction. In addition to dimensional stability and compliance with manufacturing tolerances, strength can be of major importance for the service life of a component.
process parameters layer structure filament	Apart from the type of plastic and the manufacturing process, the strength of the finished components is essentially dependent on the set process parameters and the layer structure. Important variable parameters in this context are the layer height and the track spacing, as well as the temperature of the plasticizing unit and of the platform. Additional preheating of the material may also be necessary in case the transferred thermal energy of the plasticizing unit is not sufficient to melt the filament. However, the delicacy and precision of detail of the process depend strongly on the quality of the melted filament or the strands produced and not on the diameter of the filament.
process parameters mechanical strength	If the layer height, track width and temperature are not adjusted correctly, the bonding of the layers and individual tracks may be inadequate. The result can be bonding defects or even larger gaps within a component, which significantly reduces its mechanical load-bearing capacity. However, this problem can also be caused by an excessively high production speed and therefore incomplete application of the strands. It is also important to avoid local evaporation during the process. This is caused by excessive moisture in the material and results in the drawing of undesirable strands, thus giving rise to strength differences.
shrinkage acrylonitrile-butadiene-styrene (ABS)	Plastics that expand under the influence of heat can also warp during cooling due to shrinkage. This has an additional negative effect on the stability of a component. Such is the case, for example, with the plastic acrylonitrile-butadiene-styrene (ABS), which nevertheless lends itself very well to 3D processes. Here, the limitations of ABS become apparent in the case of very complex geometries. The production personnel at the machine must take this into account accordingly during the production process.

15.4 Examples and Products

The range of applications for additive manufacturing is diverse. Additive manufacturing can be used to produce components in almost all areas (e.g., medical technology, automotive technology, aerospace technology, but also for the production of jewelry and toys). One of the advantages of 3D manufacturing processes is the production of complex geometries. (Figure 15.5) shows some application examples taken from the engineering sciences.

parts from 3D manufacturing applications

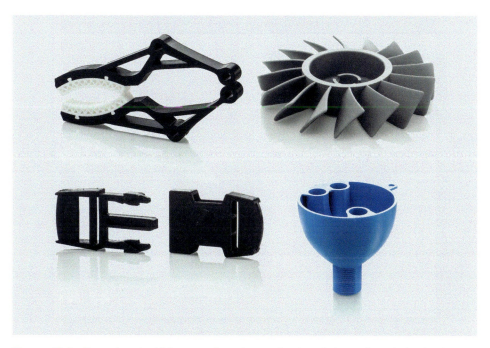

Figure 15.5 Examples of additive manufacturing applications (gripper, fan, closing buckle, funnel functional part) (source: Ultimaker B. V.)

15.5 Performance Review – Lesson 15

No.	Question	Answer Choices
15.1	To produce a component, material is _____ layer by layer in additive manufacturing (AM).	removed added milled
15.2	Additive manufacturing processes are also referred to as "rapid technologies" because they are particularly _____ compared to other production processes.	accurate continuous fast
15.3	Another name for additive manufacturing is _____.	plastics manufacturing subtractive manufacturing generative manufacturing
15.4	Because a 3D printer applies thin layers on top of each other, finished parts often exhibit a _____ surface.	very smooth stair-stepping detailed
15.5	For the finished part to ensure adequate strength, the individual layers should be _____.	well bonded uniformly thick as thin as possible
15.6	Additive technologies can only produce complex geometries with undercuts by using _____.	solvents support structures preheating
15.7	Due to anisotropy, the mechanical properties of additively manufactured components depend on the _____.	build-up direction number of layers printing speed
15.8	In addition to engineering plastics, additive manufacturing uses _____ in particular.	commodity plastics high-performance plastics

16 Lesson
Plastic Welding

Subject Area	From Plastic to Product
Key Questions	What is plastic welding?
	What kind of plastics can be welded?
	What are the technical processes for welding plastics?
Contents	16.1 Fundamentals
	16.2 Process Steps
	16.3 Welding Processes
	16.4 Examples and Products
	16.5 Performance Review – Lesson 16
Prerequisite Knowledge	Classification of Plastics (Lesson 3)
	Deformation Behavior of Plastics (Lesson 4)

■ 16.1 Fundamentals

Welding of plastics is the joining of two parts made of the same or very similar plastic under heat and pressure. The joining surfaces, also called "mating surfaces", are brought into a *thermoplastic*, i.e., molten, *state* for welding. The surfaces are then joined together under pressure and the joint is cooled until it is dimensionally stable. *definition*

The fact that the joint surfaces must be molten indicates that neither elastomers nor thermosets, but only thermoplastics can be welded. *thermoplastics*

■ 16.2 Process Steps

energy supply
welding processes

For the thermoplastic to become molten, energy must be supplied to it. Four methods are available for this purpose, which employ different physical processes. The welding processes are classified according to these methods (Figure 16.1). The table lists both the welding processes used mainly in industrial series production and those used more for manual welding. Some of the processes will be discussed in the next section.

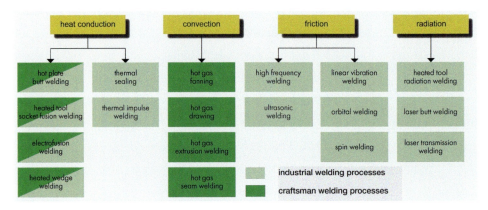

Figure 16.1 Classification of welding methods

pressure
heating time

Pressure is a very important factor besides the energy supply in the contact zone. It causes the melt to flow and the two surfaces to become permanently joined together. To ensure that the material can also mix intimately, sufficient plastic material needs to be melted. Therefore, the heating time is extremely important as well.

Welding generally consists of five steps:

process steps

- Step I: cleaning of the joining surface (removal of the oxide layer)
- Step II: heating of the joining surfaces
- Step III: applying the pressure
- Step IV: cooling down under pressure
- Step V: post-processing of the weld line

temperature range
viscosity

The weldability of two thermoplastics depends on two factors: the temperature range in which they become molten must be similar, and the viscosities of the melts must be similar. These two conditions are important for ensuring that the plastics become molten at the same time and that they can flow easily into each other to form a strong bond.

16.3 Welding Processes

Heated Tool Welding (*Heat Conduction*)

A common feature of all "heated tool welding processes" is that the heat is supplied to the joining surfaces by elements. These mostly electrically heated, metallic elements transfer the heat to the plastic by means of *heat conduction*. A basic distinction is made between direct and indirect heated tool welding.

process

In the "direct process", the heat flows straight from the heating element into the mating surface; in the "indirect process", the heat is transported from the outside through the rest of the joint part to the mating surface. Due to the poor thermal conductivity of plastics, the indirect process is only used for very thin walls (films).

direct

For illustration purposes, two processes are described here:

indirect

- *direct* hot plate butt welding
- *indirect* thermal impulse welding

"Hot plate butt welding" is a frequently used welding process for plastics. It is used, for example, to join PP and PE pipes. It is also used as an automated process for the production of automotive taillights. The welding process is shown in Figure 16.2.

hot plate butt welding
direct heated tool welding

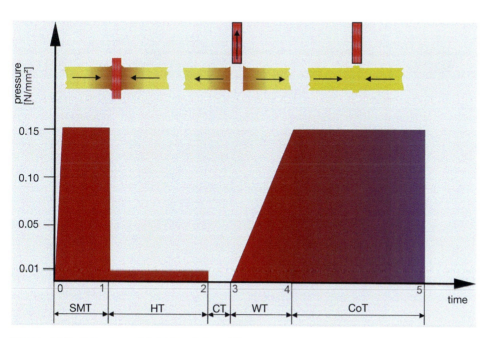

Figure 16.2 Sequence of hot plate butt welding

surface melting time (SMT)

Surface melting time (SMT): The surfaces to be joined are adapted to each other by melting. The pressure of approx. 0.15 N/mm² is applied until a closed bead is visible on the circumference of the joining surface.

heating time (HT)

Heating time (HT): The contact pressure is reduced to approx. 0.01 N/mm². At this reduced contact pressure, the surfaces are melted for welding by the heating element.

changeover time (CT)

Changeover time (CT): The heating device is pulled out as quickly as possible.

welding time (WT)
cooling time (CoT)

Welding time (WT) and *cooling time* (CoT): The surfaces to be joined are moved together until immediately before contact. Then the joining pressure is built up. This pressure must be built up quickly and evenly from 0 to the maximum value. During this process, the bead must be observed, as it has to be uniform and round. The pressure is maintained until the fusion zone is only warm to the touch.

thermal impulse welding
indirect heated tool welding

This is the most widely used *indirect* heated tool welding process. Because of the poor thermal conductivity of plastics, "thermal impulse welding" is only used for very thin films. Its main application is in packaging, for sealing bags, pouches, and sacks. The process is shown in Figure 16.3.

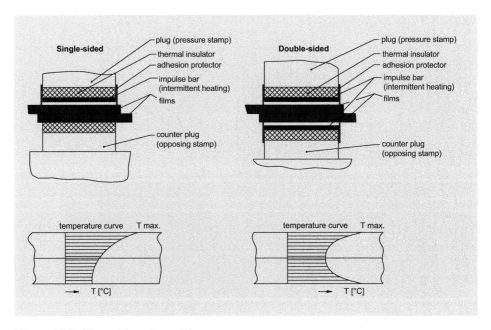

Figure 16.3 Thermal impulse welding

thermal conduction

Welding is performed by heating thin metal rails with a high current pulse. These bars are provided with anti-adhesion coatings and transfer the heat by thermal conduction to the films. The films melt and weld together. Both one-sided and two-sided processes exist. In the single-sided method, the films are heated by a metal bar from one side only; in the two-sided method, the films are heated from both sides.

As a result of the processes, there is an unfavorable heat distribution within the parts to be welded. It is necessary to attain the melting temperature of the contact point of the films without allowing the warmer (outer) edge to reach the decomposition temperature of the plastic.

heat distribution

Hot Gas Welding (*Convection*)

"Hot gas welding processes" are another group of welding processes. They are mostly performed by hand and require a high degree of manual skill. Here, hot gas, e.g., clean compressed air, is used for heating. The joint surfaces are heated with the hot gas and welded under pressure, usually with an additional material. The process is mainly used for the manufacture and repair of components in equipment, tank, landfill, and pipeline construction.

hot gas welding

A distinction is made between *hot gas fanning, hot gas drawing,* and *hot gas extrusion* welding. In hot gas fanning and hot gas drawing welding, a plastic filler rod serves as a round or profiled rod. The joining surfaces of the base material and the filler rod are plasticized by means of hot gas and joined under pressure. In hot gas fan welding (Figure 16.4) the filler rod is guided freely, and the joining surfaces are plasticized in a fanning movement. The manual process is also known as "permanent hot gas welding".

hot gas fanning
hot gas drawing
welding filler rod

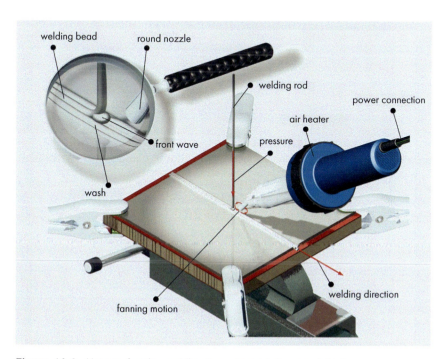

Figure 16.4 Hot gas fanning welding (permanent hot gas welding)

In *hot gas drawing welding* or *high-speed welding* (Figure 16.5), the filler rod is guided in a drawing nozzle and introduced into the weld joint under pressure. The device is also called a "speed gun".

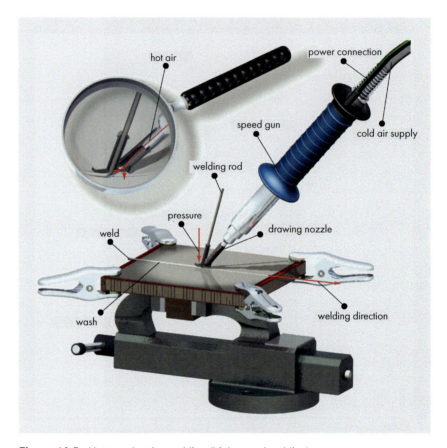

Figure 16.5 Hot gas drawing welding (high speed welding)

Hot gas drawing welding is the faster and more uniform welding process and is usually preferred to hot gas fan welding.

hot gas extrusion welding
welding shoe

In "hot gas extrusion welding", the continuous welding process (Figure 16.6) is primarily used. Here, a filler metal is usually fed to the plasticizing unit in the form of a wire, melted in the extruder and fed to the welding shoe as a plasticized filler metal for the weld joint. The correspondingly manufactured welding shoe forms the weld seam geometry. The weld zone is preheated by a hot gas blower attached to the welding unit. This welding process is a fast welding process with a high mass output.

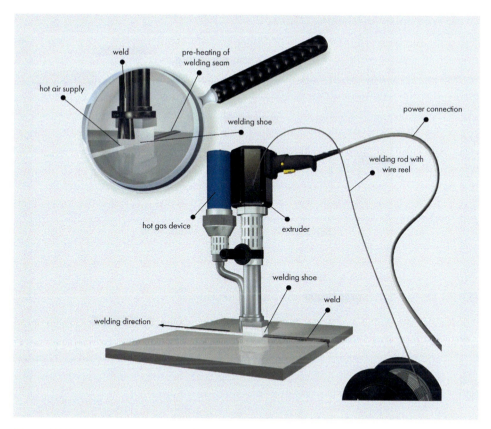

Figure 16.6 hot gas extrusion welding

Friction Welding (*Friction*)

In the "friction welding process", frictional heat is employed to melt the plastic. Examples given here are *spin welding* and *ultrasonic welding*.

friction heat

In "spin welding", rotationally symmetrical parts are welded by "external friction". While one part is rotating, the other is stationary and is pressed against the rotating one under a certain force.

external friction
spin welding

The mating surfaces adapt to each other by melting. After a sufficiently large weld bead has been generated at the seam, the clamping fixture is released, and the seam cools down under pressure.

In "ultrasonic welding", the material is melted by "internal friction". In this process, the mechanical damping capacity of the plastic is utilized. A device is used to generate high-frequency mechanical vibration. This vibration passes through the workpiece and is reflected at the anvil, thus creating a standing wave. If the damping of the workpiece is too high, it absorbs the vibration and it cannot reach the joining surface. The process is used in large-scale manufacturing for household goods, electronics, and toys.

ultrasonic welding
internal friction

Laser Transmission Welding (*Radiation*)

The still relatively young process of "laser transmission welding" of plastics has certain advantages when compared to other welding processes. These include a very small area affected by heat, low melt expulsion, feasibility of 3D seam geometries, flexible welding system technology, especially for small and medium quantities, as well as the welding of thermoplastic elastomers and the joining of micro components. Wherever small melt cushions and low melt expulsion are to be achieved, laser beam welding is suitable. This may lead to substitution of adhesive bonding technology. Furthermore, components with inserted electronic or micromechanical components can be welded without damage. This *radiation joining process* is becoming particularly interesting for industrial applications due to low-cost high-power diode lasers.

reduced heat-affected zone
laser transmission welding

■ 16.4 Examples and Products

As a joining process, welding is just as important for plastics as it is for the metals steel and aluminum. The most important criterion for all products joined by welding is the quality of the weld itself. The strength of the entire component must not be negatively affected by the weld or, if so, only to a minor extent.

Hot plate butt welding is used for joining PP and PE pipes (Figure 16.7).

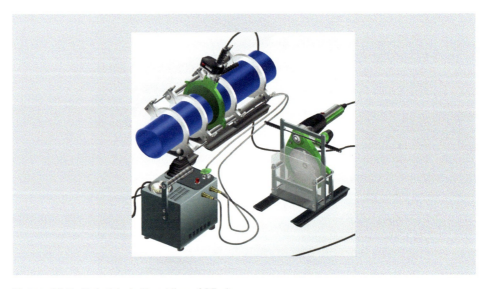

Figure 16.7 Hot plate butt welding of PE pipes

PE pipes have been used in gas and water supply for more than 60 years.

Another important area for welding applications is the manufacturing of large apparatus and vessels by hot gas welding (Figure 16.8).

Figure 16.8 *Pickling tank* welded from PP sheet material, *mixing tank* welded from PP semi-finished material with dosing station, *distribution pipes for nozzle sections* welded from PP pipe for a washer of a coal-fired power plant (source: STEULER-KCH GmbH, Höhr-Grenzhausen)

■ 16.5 Performance Review – Lesson 16

No.	Question	Answer Choices
16.1	The mating surfaces of the plastic parts become _____ during the welding process.	thermoplastic thermoelastic
16.2	Thermosets and elastomers _____ be welded.	can cannot
16.3	Two different plastics can be welded together provided they have a similar _____ and similar _____.	thermal diffusivity viscosity melting temperature color
16.4	In hot plate welding, the heat is conveyed to the plastic through _____.	convection radiation heat conduction
16.5	_____ is often used for the repair of containers.	Hot gas welding Hot plate welding Induction welding

No.	Question	Answer Choices
16.6	In hot gas extrusion welding, the weld geometry is formed by the _____.	welding rod welding shoe welding pressure
16.7	_____ is a welding process with a very low heat-affected zone.	hot gas welding hot plate butt welding laser transmission welding

17 Lesson
Machining Plastics

Subject Area	From Plastic to Product
Key Questions	What properties of plastics affect mechanical machining?
	What machining rules result from these effects?
	What machining methods and tools are employed?
Contents	17.1 Fundamentals
	17.2 Cutting Processes
	17.3 Performance Review – Lesson 17
Prerequisite Knowledge	Classification of Plastics (Lesson 3)
	Deformation Behavior of Plastics (Lesson 4)

■ 17.1 Fundamentals

The machining processes used in plastics processing include sawing, milling, turning, drilling, grinding, and polishing. *methods*

The experience gained in using these processes for metal machining cannot be directly transferred to plastics machining. This is because plastics display properties different from those of metal. *plastics properties*

- Plastic conducts heat more poorly than metal. Therefore, the heat generated by friction during machining will not be dissipated well by the material. The interface must thus be cooled particularly well to prevent the plastic from melting or even decomposing.
- The thermal expansion of plastics is very high. Consequently, the saw blade may jam when cutting through plastic. When being drilled, the plastic may

assume different dimensions, for example: after cooling, the holes may be 0.05 to 0.1 mm (2–4 mils) smaller than the selected drill bit.

- Plastics are particularly susceptible to notching. The cuts must be smooth when machining them to avoid a reduction in mechanical load-bearing capacity.
- Plastics generally exhibit less strength than metals. Machining therefore requires significantly lower machining forces.

machining principles

The above properties result in rules that must be respected during machining:

- Thermoplastics should not heat up above 60 °C (140 °F) during processing and thermosets should not heat up above 150 °C (302 °F).
- The heating can be influenced by the cutting speed, the feed rate, and the tool geometry. Furthermore, it is possible to cool the cutting point down by means of cooling media.
- For the creation of even cuts, smooth-running machines should be used.

■ 17.2 Cutting Processes

Sawing

circular saw

Circular saws use high-speed steel or carbide-tipped blades, which must be hollow ground. The tooth pitch should be relatively small (Figure 17.1).

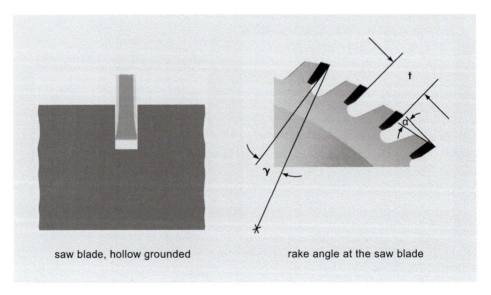

Figure 17.1 Design of saw blade teeth

On band saws, the teeth are slightly interlaced to avoid smearing the tooth gaps with plastic. Some reference values for sawing plastics are given in Table 17.1:

Table 17.1 Reference Values for Sawing Plastics

Plastics	Sawing tool	α (°)	γ (°)	t (°)
thermoplastics	SS (high speed steel)	30–40	5–8	2–8
	HM (carbide)	10–15	0–5	2–8
thermosets	SS	30–40	5–8	4–8
	HM	10–15	3–8	8–18

Milling

Plastic milling cutters have a lower number of cutting edges compared to metal milling cutters, but a higher number compared to wood milling cutters. They consist of high-speed steel or carbide but can also be carbide-tipped. The cutting speed should be kept as high as possible and at a relatively low feed rate. The harder the plastic material, the smaller the rake angle should be (Figure 17.2).

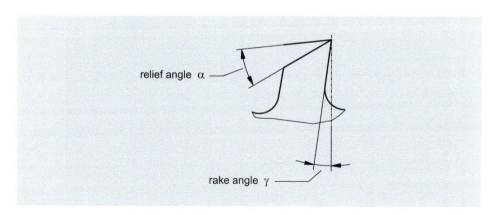

Figure 17.2 Tool angles for milling cutters

The softer the material, the smaller the number of cutting edges and the higher the feed rate that should be selected. Some reference values for milling of plastics are given in Table 17.2.

Table 17.2 Reference Values for Milling Plastics

Plastics	Milling tool	α (°)	γ (°)
thermoplastics	SS (high speed steel)	2–15	up to 15
thermosets	SS (high speed steel)	up to 15	15–25
	HM (carbide)	up to 10	5–15

Drilling

Twist drills for metallic materials can also be used for plastics. Drills with steep helix are better for removing the chips.

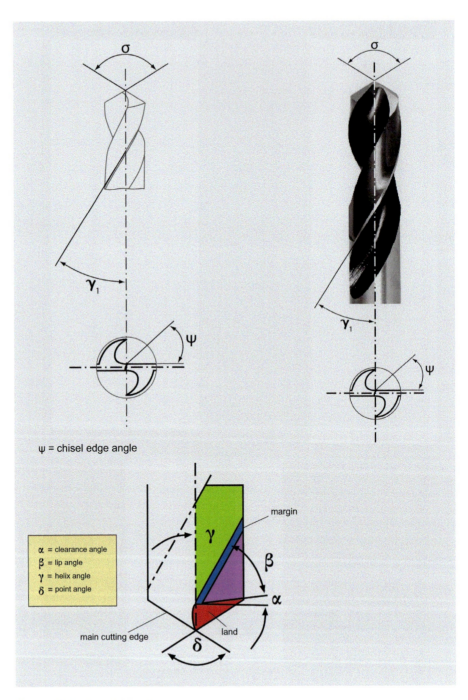

Figure 17.3 Twist drill angles

Because of the frictional heat which is generated during drilling and which causes the plastics to expand significantly, the holes are 0.05 to 0.1 mm (2–4 mils) smaller than the diameter of the drill bit. In practice, this is allowed for by selecting a larger drill to obtain the desired dimension.

frictional heat

For slightly smearing materials, such as PE and PP, a high feed rate is used at a low cutting speed in order to dissipate the heat along with the chips. For drill diameters of 10 to 150 mm (0.4–6.0 in), a diamond-tipped hollow drill is used. Several reference values for drilling plastics are given in Table 17.3:

heat dissipation

Table 17.3 Reference Values for Drilling Plastics

Plastics	Drill tool	α (°)	γ (°)	φ (°)
thermoplastics	SS (high speed steel)	3–12	3–5	60–110
thermosets	SS (high speed steel)	6–8	6–10	100–120
	HM (carbide)	6–8	6–10	100–120

Turning

The lathe should operate at high speeds and be equipped with a liquid cooling system. The turning steels may be of high speed steel, depending on the plastic. The designation of the tool angles of turning steels for plastics is given in Figure 17.4.

turning tools

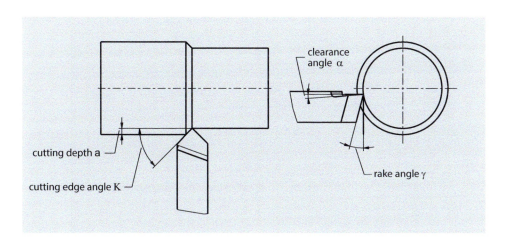

Figure 17.4 Tool angles for lathe steels

For thermosets and plastics with glass fiber fillings, turning tools with carbide cutting edges are used. Some reference values for turning plastics are given in Table 17.4:

Table 17.4 Reference Values for Turning Plastics

Plastics	Turning tool	α (°)	γ (°)	κ (°)	a
thermoplastics	SS (high speed steel)	5–15	up to 10	15–60	up to 6
thermosets	SS (high speed steel)	5–10	15–25	45–60	up to 5
	HM (carbide)		10–15	45–60	up to 5

Grinding and Polishing

grinding

Grinding is performed with commercially available abrasive papers or with abrasive belts. The grinding speed of the belts should be approx. 10 m/s.

polishing

Felt or buffing wheels with polishing agent are used for polishing plastics. In order not to melt the surface of thermoplastics during polishing, the process is interrupted repeatedly.

■ 17.3 Performance Review – Lesson 17

No.	Question	Answer Choices
17.1	The plastic can melt at the interface during sawing, as it conducts heat _____ efficiently than metal.	more / less
17.2	The saw blade can jam in the plastic during cutting due to the high _____ of the plastic.	thermal conductivity / thermal expansion / viscosity
17.3	After cooling, the diameter of a hole drilled into plastic is _____ the diameter of the drill bit.	larger than / smaller than / exactly the same size as
17.4	The same drill bits _____ used for both plastics and metals.	can / cannot
17.5	Milling should be performed at the _____ possible cutting speed.	highest / lowest
17.6	For machining plastics, the lathe should be equipped with a _____ system.	liquid cooling / air cooling
17.7	When polishing plastic surfaces, the process _____ be interrupted several times, because the resulting frictional heat could melt the surface.	must / must not

18 Lesson
Bonding of Plastics

Subject Area	From Plastic to Product
Key Questions	What are the physical and chemical principles on which bonding is based?
	What bonding methods exist?
	How should an adhesive bond be formed?
	Which plastics can be bonded to one another?
Contents	18.1 Fundamentals
	18.2 Bondability and Adhesive Joints
	18.3 Classification of Adhesives
	18.4 Bonding Process
	18.5 Examples and Products
	18.6 Performance Review – Lesson 18
Prerequisite Knowledge	Classification of Plastics (Lesson 3)

■ 18.1 Fundamentals

The bonding of plastics serves as a full surface joining technique. Unlike the case for welding, all types of plastics, i.e., also elastomers and thermosets, can be bonded. In addition, even plastics of very different types can be bonded to each other and to other materials. The advantages of adhesive bonding include:

joining technology

- It is possible to bond thin, small and complicated parts.
- Ideal for sealing, vibration damping, thermal and electrical insulation, and for bridging uneven surfaces.
- Highly aesthetic solutions can be designed.
- Bonded joints achieve comparable strengths to other processes (e.g., welding).

18 Bonding of Plastics

bonding mechanism The mechanism of bonding is based on the internal bond of the adhesive, "the cohesion", and on the bond between the adhesive and the part to be bonded, "the adhesion" (Figure 18.1).

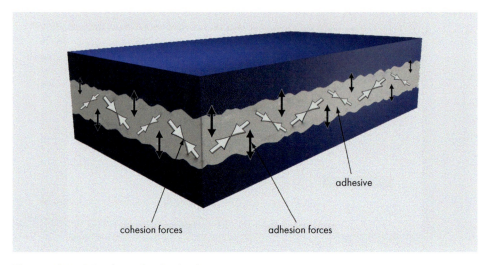

Figure 18.1 Adhesive and cohesive forces

adhesion The adhesive forces are very small and only apply when there is direct contact between the adhesive and the surface. Therefore, nothing must interfere with this contact. To ensure this, the surface to be bonded must be cleaned of grease and dirt particles before applying the adhesive.

cohesion Generally, cohesion only acts within the adhesive. However, a cohesive force can also occur between the two parts to be joined if the plastic to be bonded is soluble.

Mechanical adhesion consists in the anchoring of the adhesive within the surface roughnesses of the part to be joined (Figure 18.2).

solvent-based bonding A pure solvent is applied to the surfaces, diffuses into them, and dissolves the plastic, breaking the intramolecular bonds between the molecules at this point. When the parts to be joined are pressed onto each other, their molecules hook together and form a strong bond due to cohesive forces (Figure 18.3).

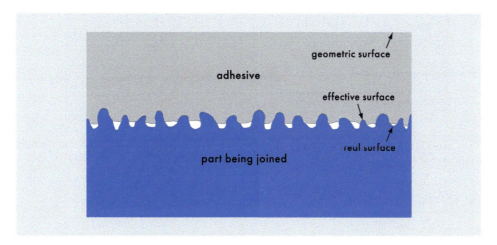

Figure 18.2 Surface roughness during bonding

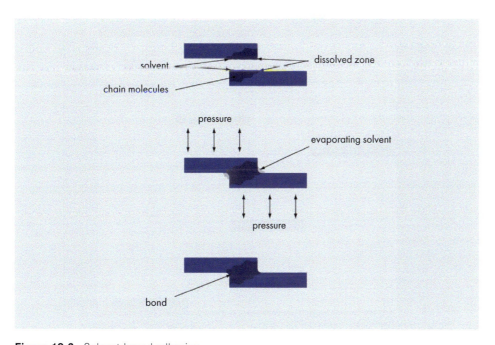

Figure 18.3 Solvent-based adhesion

18.2 Bondability and Adhesive Joints

bondability

In addition to the degree of cleanliness, roughness and solubility of the joining surfaces, their polarity and wettability also play a decisive role in whether two parts can be bonded together easily (Figure 18.4).

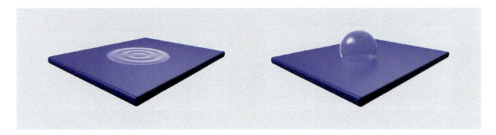

Figure 18.4 Good and poor wetting of surfaces

Not all plastics are equally suitable for high-quality adhesive bonding. They differ in terms of wettability, polarity, solubility and bondability. An overview of the bondability of different plastics is given in Table 18.1:

Table 18.1 Bondability of Plastics

Plastic	Wettability	Polarity	Solubility	Bondability
polyethylene (PE)	–	nonpolar	insoluble	–
polycarbonate (PC)	+	+	+	++
polystyrene (PS)	–	nonpolar	soluble	++
rigid polyvinyl chloride (PVC)	++	polar	soluble	++
polymethyl methacrylate (PMMA)	++	polar	soluble	++
phenol formaldehyde resin (PF)	++	polar	insoluble	++
unsaturated polyester resin (UP)	++	polar	insoluble	++
polyamide 66	++	polar	hard to dissolve	+

good: ++   fair: +   poor: –

types of stresses

In addition to the nature of the actual bonding surfaces, their type and position on the parts to be bonded play an important role. The different types of stresses that the forces in a bonded seam can cause can be seen in Figure 18.5.

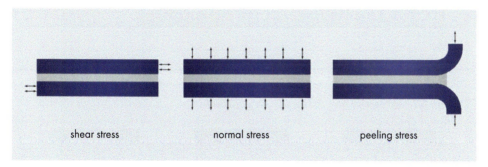

Figure 18.5 Types of stresses on an adhesive joint

The force that later acts on the seam should, if possible, give rise to shear stress in the seam and not a peeling effect. A peeling effect reduces the load-bearing capacity of a bonded joint.

peeling stress

Some possible types of bonded joints are shown in Figure 18.6.

types of joints

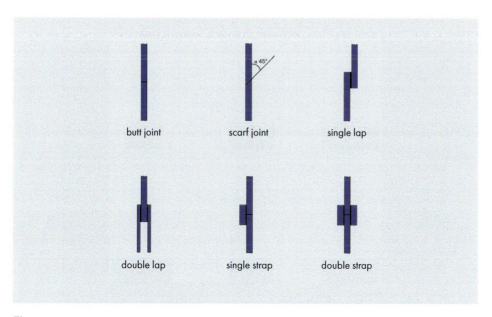

Figure 18.6 Types of adhesive joints

The forces that a joint can absorb and thus the forces that it can transmit are largely proportional to the surface area. A "scarf joint" at 45° can thus transmit an approx. 40% greater force than a "butt joint".

18.3 Classification of Adhesives

Adhesives can be divided into physically and chemically setting adhesives.

Physically Setting Adhesives

solvents
solvent effects

To achieve good wetting of the surfaces to be bonded, adhesives are often dissolved in an organic solvent or dispersed in water (very finely dispersed). For the adhesive to become solid and the joining surfaces to bond firmly, the solvent must be able to escape from the adhesive. Either it evaporates or it is absorbed by the joining surface. Prior to this, it is necessary to clarify whether or not the solvent has a negative effect on the plastic. For example, it might release internal stresses in the plastic, which could lead to stress cracks in the part.

contact adhesives

An example of a solvent-based adhesive is a contact adhesive. In this case, the parts to be joined that have been wetted with adhesive must remain open until the solvent has evaporated from the adhesive. Only when the adhesive feels dry can the surfaces be pressed together and bonded. Correction of the bond is no longer possible here.

solvent-based bonding

This procedure was mentioned earlier in this chapter (see Figure 18.3). Solvents melt the surfaces of the plastics such that they intermingle, leaving a welded joint when the solvent evaporates. Solvent-based bonding is also referred to as "solvent sealing", "solvent welding" or "solvent cementing".

hotmelt adhesives

Hotmelt adhesives are applied as a plasticized mass to the joining surfaces, which are then pressed together. The adhesive becomes solid when it cools down. Since this cooling time is very short, this process is often used in large-scale production.

Chemically Setting Adhesives (Reaction Adhesives)

reaction adhesives

Reaction adhesives, as the name implies, bond through a chemical reaction. This reaction, which can be polymerization, polyaddition or polycondensation, results in crosslinked macromolecules (thermosets). Depending on the system, the reaction is initiated by hardener, accelerator, or heat.

curing

The adhesive can be mixed from the various components (two- or multi-component systems) only just before processing, as the reaction starts quickly and the ready-mixed adhesive cures. After curing, it can no longer be processed.

18.4 Bonding Process

The quality of the bonded joint is decisively influenced by the execution of the bonding process. As indicated before, the bonding takes place in the following steps:

Preparation of Suitable Bonding Surfaces

The most important requirement for bonding is that the parts to be bonded and the bonded seams are designed for bonding. This design determines the type of stress that an applied force exerts on the bonded joint. As already mentioned, an applied force should not have a peeling effect on the bond, if possible.

quality

bonding surfaces

Cleaning and Degreasing of the Bonding Surfaces

Furthermore, it is important that the bond is not impaired by contamination. Depending on the contamination, cleaning is done in continuous baths with organic solvents or alkaline cleaning agents, ultrasonic baths or vapor degreasing baths.

cleaning

Pretreatment of the Bonding Surfaces

To further enhance the surface properties for bonding, they must be pretreated. For plastics that are easy to bond, this can be done by mechanical abrasive methods (grinding, sandblasting) or chemical roughening (etching). Surfaces of plastics that are difficult to bond are activated by the addition of oxygen (flame treatment, corona treatment) or coating.

pretreatment

Application of the Adhesive

During application of the adhesive, uniform wetting of the joining surfaces and a constant layer thickness are essential.

application

Waiting until the Adhesive is Ready to Bond

The time that must elapse before the adhesive is ready for bonding varies greatly from adhesive to adhesive. It must be strictly adhered to, as otherwise the bonding mechanism will be impaired and either no bond at all or only a poor bond will be formed.

time

Joining and Fixing the Parts to Be Bonded

After the parts to be bonded have been joined, pressure is applied; this displaces air between the joining surfaces and thus also determines the adhesive film thickness. For some adhesives that have a longer curing time, it is useful to fix the parts to be bonded after pressing them together to prevent them from displacing.

joining

Curing the Adhesive

curing

The various adhesives have individual curing times. The curing time must always be observed before the bonded joint can be loaded.

Removing the Fixation from the Glued Parts

fixation

When the adhesive has cured, the fixture can be removed from the bonded parts. However, although the adhesive has cured to the point where the parts can no longer move, it is often necessary to wait a further time before the bonded joint can withstand full loads.

When carried out correctly, adhesive bonding of plastics can now be regarded as a totally adequate joining process that can be used to produce inseparable joints with a high degree of traction.

■ 18.5 Examples and Products

Figure 18.7 below shows applications for bonded pipes.

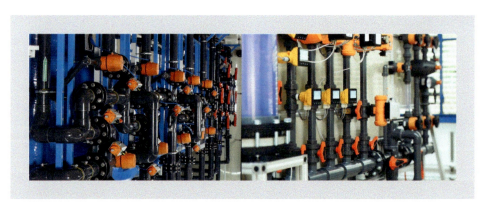

Figure 18.7 Applications in the field of water treatment for bonded pipes (source: Georg Fischer Piping Systems AG)

18.6 Performance Review – Lesson 18

No.	Question	Answer Choices
18.1	Unlike welding technology, adhesive bonding techniques _____ be applied to elastomers and thermosets.	can cannot
18.2	The mechanism of adhesive bonding is based on _____.	cohesion adhesion cohesion and adhesion
18.3	In order for the bond to have excellent strength, the surfaces to be joined must be especially _____.	large smooth clean
18.4	The forces acting on the adhesive joint should not exert _____.	tensile pressure peeling
18.5	After curing, reaction adhesives become _____.	thermosets elastomers thermoplastics
18.6	Reaction adhesives are only mixed shortly before processing, as they react _____ and can then no longer be processed.	quickly slowly
18.7	A "scarf joint" transmits a _____ force than a "butt joint".	bigger smaller
18.8	_____ joints may be produced by means of adhesive bonding.	Only low-quality Even high-quality

19 Lesson
Plastic Waste

Subject Area	Ecology of Plastics
Key Questions	How much plastic is produced?
	Which products are made of plastic?
	What is the lifetime of plastic products?
	What specific problems does plastic waste entail?
	What are the advantages of recycling plastics?
	What must waste management consider when handling plastic waste?
Contents	19.1 Fundamentals
	19.2 Plastics Production and Its Applications
	19.3 Plastic Products and Lifetime
	19.4 Prevention and Recycling of Plastic Waste
	19.5 Circular Economy in Plastics Business
	19.6 Performance Review – Lesson 19
Prerequisite Knowledge	Classification of Plastics (Lesson 3)
	Deformation Behavior of Plastics (Lesson 4)

■ 19.1 Fundamentals

In recent years, plastic waste has increasingly become the focus of criticism. The impact of plastic waste can mainly be summarized in four points:

- Relative to its weight, plastic waste has a large volume and is difficult to compress.
- Plastic waste is generally poorly degradable, which means that it cannot be incorporated into the biological cycle.

plastic waste

volume issue

degradability

harmful substances

- Plastic waste sometimes contains ingredients that cause problems when combusted in waste incineration plants. These are, for example, chlorine in PVC, nitrogen in PUR and PA, fluorine in PTFE, sulfur in plastic rubber and heavy metal additives in many plastics.

recyclability

- The recyclability of plastic waste depends on the grade purity as well as the degree of contamination.

prevention and recycling

All these problems would recede if it were possible to avoid and recycle plastic waste more successfully.

Let's take a closer look at the quantity and composition of plastic waste below.

■ 19.2 Plastics Production and Its Applications

plastic production

Global plastics production has grown steadily in recent decades. For example, it has increased more than sevenfold since 1976, that is, over the last 45 years. In the same period, the world's population has grown from just over 4.2 billion people to over 7.6 billion, an increase of only about 80%. The plastics market, and thus global plastics production, is therefore growing at a considerably above-average rate. Germany produces more plastic and plastic products than it consumes itself, so that exports are greater than imports. Figure 19.1 shows global and European plastics production since 1950.

The amount of plastics produced worldwide in 2019 was slightly less than 370 million tons (Mt). Comparing the European production values with the global values, Europe has shown a certain stagnation for almost 20 years at the level of 60 million tons. The drivers of the global increase in the production of plastics are distinctly the Asian countries, in particular China, which in recent years has recorded enormous economic growth rates in addition to a strong increase in population.

annual consumption

In Germany, almost 20 million tons of plastics were produced in 2019. The largest percentage was accounted for by the plastic PE, which is used particularly in the packaging sector. The dominance of the packaging sector is also reflected in the consumption of plastics, shown in Figure 19.2 according to individual areas of application. These figures include recycled plastics.

19.2 Plastics Production and Its Applications

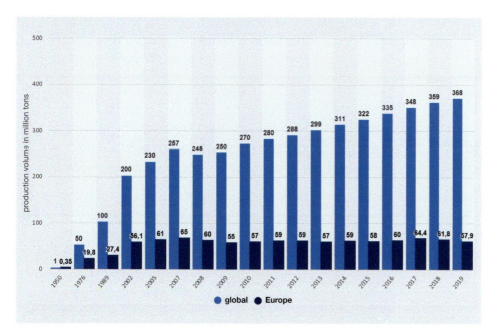

annual production

Figure 19.1 Global and European plastics production from 1950 to 2019 (in million tons) (source: PlasticsEurope)

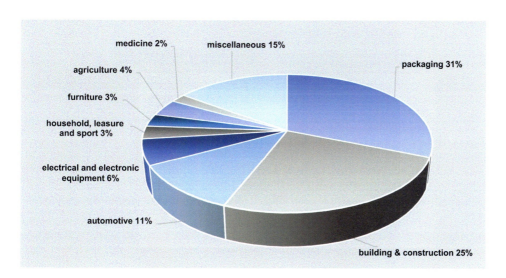

Figure 19.2 Use of plastics (virgin and recycled) in % for 2019 (source: PlasticsEurope)

Besides the packaging sector, the construction sector (e.g., for building insulation) is playing an increasingly significant role in the use of plastics. More than 50% of the total volume of plastics in Germany is processed in this sector. As stated above, in 2019 PE (PE-LD, PE-HD) has the largest share in terms of individual plastic types at 27%, followed by PP at just under 20% and PS at 5%. Polyvinyl chloride (PVC), which is widely used in the construction sector and is used to make plastic windows, shutters, doors, and pipes for underground applications, also has a notable share of 15%.

■ 19.3 Plastic Products and Lifetime

lifetime

The lifetime of plastic products is usually underestimated. In the public mind, plastic is still mainly associated with disposable articles. The reason for this is certainly the use of plastics for packaging, especially disposable packaging.

packaging

However, if we take a closer look at the different applications of plastic products that we learned about in the previous section, we see that packaging accounts for about one-third of total consumption. Nevertheless, most applications are those in which plastic is used for long-lasting products due to its high performance. The following studies show the distribution of the lifetimes of plastic products (Figure 19.3).

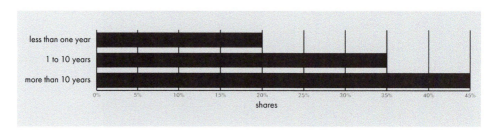

Figure 19.3 Lifetime of plastic products

usage time

As we can see here, about 20% of plastic products are disposed of within one year (which includes, for example, packaging materials or disposable bottles), while 35% of all plastic products are in use for a period of one to 10 years; 45% of plastic products only become waste after more than 10 years. An example of durable plastic products are plastic windows, which have a lifespan of more than 50 years. Plastic pipes made of polyethylene for gas and water supply, which are embedded in the ground, have a lifespan of 100 years. Thus, many plastic components will not be waste until many years after they have been produced.

optical storage media

For example, optical storage media such as the CD are long-life-cycle articles, and the CD box will also generally only be generated as waste after a long time. In contrast, the protective or packaging film in which the CD box is wrapped will be recyclable plastic immediately after purchase. This, like most packaging, is a short-life product, albeit one which is recyclable.

In 2019, the total amount of plastic consumed in the Federal Republic of Germany resulted in approximately 6.3 million tons of plastic waste, including scrap from production and processing. Some 5.35 million tons thereof was accounted for by the so-called post-consumer sector, i.e., the plastic waste generated by end users or consumers (Table 19.1).

generated waste

Table 19.1 Recovery of Plastic Waste in Germany for 2019 (source: Federal Environment Agency 2020, internal compilation with data from the CONVERSIO Market & Strategy GmbH)

Recovery method	Total plastic waste *	Post-consumer waste **
material recycling	2.93 million tons	2.06 million tons
▪ mechanical recycling	2.92 million tons	2.05 million tons
▪ chemical recycling	0.01 million tons	0.01 million tons
energy recovery	3.31 million tons	3.25 million tons
▪ waste incineration plants	2.15 million tons	2.12 million tons
▪ fuel/other	1.16 million tons	1.13 million tons
disposal/landfilling	0.04 million tons	0.03 million tons
total waste generated	6.28 million tons	5.35 million tons

* including plastic waste from production and processing
** plastic waste after use/end consumer waste except production and processing waste

The volume of plastic waste will continue to increase significantly in the coming years. The reason for this can be found in the long-life products, which will then account for around two thirds of the waste. It is encouraging to note that only less than 1% of the plastic waste collected under the Circular Economy Act ends up in a landfill. About half of the total plastic waste is recycled and the other half is recovered for energy.

domestic waste

It is evident that the recycling rate from waste from plastics production and processing is significantly better than that from the post-consumer sector. The reason for this is that in industry the waste is sorted and clean, but in households and commercial enterprises it is contaminated and mixed and must undergo further treatment before it can be recycled.

■ 19.4 Avoiding and Recycling Plastic Waste

waste fraction
blue, yellow, green bin

The two terms "waste prevention" and "waste recycling" are repeatedly cited as possible solutions in discussions on waste problems. Both topics are important. In Germany, waste recycling is carried out, for example, by sorting household waste into different waste fractions (blue, yellow, and green garbage bins).

waste prevention

Waste prevention, which is preferable to waste recycling according to the Circular Economy Act, aims to reduce the amount of waste and pollution directly during production. This can be achieved, for example, through multiple use or reuse of products. The deposit-returnable bottle is a good practical example. All in all, waste avoidance means departing from a throwaway society and moving toward a responsible and exhaustive use of long-lasting products that are truly needed. Products that are used repeatedly for a long time do not generate waste until much later. This means that far less waste would have to be recycled or disposed of. The waste problem can therefore be tackled at the roots, i.e., mass production and the unnecessarily high level of product consumption.

waste recycling

But consideration to waste recycling should also be given as early as the production stage. For this purpose, the products should be designed in such a way that they are generally recyclable and that they can be collected by type. In distribution, precautions should also be taken to ensure that the products can be brought in for recycling after use.

examples

Below are a few examples of how the prevention and recycling of waste could be considered over the course of production:

- Injection molds with hot runner systems can avoid large amounts of production waste. A "hot runner" is a heated part of the sprue in an injection mold. Since this channel is heated, the plastic melt inside cannot solidify and subsequently can be used for the next shot. This saves large quantities of sprue waste that would otherwise have to be recycled or disposed of.

- Some additives contain toxic heavy metals. When these kinds of additive can be replaced by less toxic substances, hazardous waste, which is generated when the waste is incinerated, is avoided.

- A lot of products consist of several different materials. To be able to separate these materials easily, the products must be specially designed. For example, products should be designed to be recyclable so that they can easily be repaired or disassembled. This is the only way to recycle them once they are defective or used up.

The first German Packaging Act was enacted in 1991. Since then, the principle of producer responsibility has applied in Germany: manufacturers and distributors must take back packaging when it has fulfilled its purpose, recycle it in an environmentally friendly manner and then document this recovery process.

Dual System

The German Packaging Act (VerpackG) is the German implementation of the European Packaging Directive 94/62/EC (PACK) regulating the distribution of packaging. It also regulates the return and recovery of packaging waste. The purpose of the uniform federal law is to prevent or reduce the impact of packaging waste on the environment.

VerpackG

The Federal Council of Germany gave its final approval to the amendment to the Packaging Act on May 28, 2021.

One of the companies assuming these duties in Germany is, for example, the "Duales System Deutschland" (DSD), a company with a market share of just under 20% (2020). Several companies of a similar size to the DSD company operate in this sector. These companies serve all players in the distribution chain – from packaging manufacturers to fillers and retailers.

DSD

In Germany, many packaging manufacturers are registered on the Central Packaging Register. They have launched new logos with packaging separation information (*www.trenn-hinweis.de*).

Plastic products that are recyclable are marked with a so-called green dot (German: "Grüner Punkt"). The revenues from the "Grüner Punkt" are used to finance the dual waste system. The percentage development of plastic waste recycling, including production and processing waste from the year 1994 to 2019, is shown in Figure 20.6 of the next lesson.

"Grüner Punkt"

■ 19.5 Circular Economy in Plastics Business

Circular economy is a model of production and consumption (see Figure 19.4), in which existing materials and products are shared, leased, reused, repaired, refurbished, and recycled for as long as possible. By this means, the life cycle of the products is thus significantly extended.

Figure 19.4 Circular economy model (source: European Parliament)

In practice, this means that waste is reduced to a minimum. After a product has reached the end of its service life, the resources and materials remain a part of the economy as far as possible. They are thus productively reused again and again so that they continue to generate economic value added.

Plastics can make a significant contribution to circular economy. Thanks to their low weight and durability, and their customizable properties, plastics can save resources in any of the applications shown in Figure 19.2. And thanks to their adjustable properties, the recycling concept can already be considered when designing product components. During manufacturing, the sorted production waste can be easily reused.

The closed loop capability of plastics can be further increased through distribution-regulated take-back concepts, through reuse and repair as well as through collection for recycling.

19.6 Performance Review – Lesson 19

No.	Question	Answer Choices
19.1	In 2019, approximately _____ million tons of plastic were produced worldwide.	120 240 370
19.2	Packaging is a _____ plastic article. Beverage containers are _____ plastic products. Plastic windows are _____ products.	long-life short-life
19.3	Plastic products are often long-lasting, so that about _____ percent only become waste after more than 10 years.	20 35 45
19.4	Compared to its weight, plastic waste consumes _____ space in landfills.	little a lot of
19.5	Plastics _____ decompose, unlike pure biological waste.	do not hardly
19.6	During waste incineration, some plastics release pollutants, so avoiding and recycling plastic waste is _____ to incineration.	preferable not preferable

20 Lesson
Plastics Recycling

Subject Area	Ecology of Plastics
Key Questions	Is plastic waste recyclable at all?
	What are the basic recycling solutions for plastic products?
	What are the technical requirements for plastic recycling?
	What is the recycling potential of plastics?
Contents	20.1 Fundamentals
	20.2 Mechanical Recycling
	20.3 Chemical Recycling
	20.4 Thermal Recovery
	20.5 Recycling Plastic Waste
	20.6 Examples and Products
	20.7 Performance Review – Lesson 20
Prerequisite Knowledge	Classification of Plastics (Lesson 3)
	Deformation Behavior of Plastics (Lesson 4)
	Plastic Waste (Lesson 19)

■ 20.1 Fundamentals

The buzzwords are well known: "climate change", "pollution of the oceans", "finiteness of raw materials". Plastics are directly or indirectly affected by many of these conditions. On one hand, their wide range of applications and adjustable material properties show that plastics can help with the conservation of resources. How-

problem of waste

ever, with the associated increase in plastics production, it is necessary on the other hand to avoid generating huge quantities of plastic waste and polluting the oceans. It is imperative to avoid waste and to subject every product to a recycling process.

recycling

Recycling can not only reduce the amount of waste, but also save raw materials and energy for the production of new material.

No other material offers as many recycling options as plastics. Which recycling process is the technically, ecologically, and economically most appropriate depends on several factors, which are discussed below.

closed loop systems
open loop systems

A distinction is made between a "closed loop" process, such as for (unpolluted) production waste, and an "open loop" process, such as for plastic windows, which are only replaced after 30–40 years of use. Closed loop recycling is when the recycled material produced is reused in the same product category. Open loop recycling is when products are reprocessed and the recycled material produced is used in a different application.[1]

The shorter the cycle is, the less effort that is usually required to recycle the plastic product. As mentioned above, which recycling process makes the most sense technically, ecologically, and economically depends on a number of factors.

recycling options

The different recycling options are shown in Figure 20.1. Here, we can see scope for both recycling and recirculation of plastics.

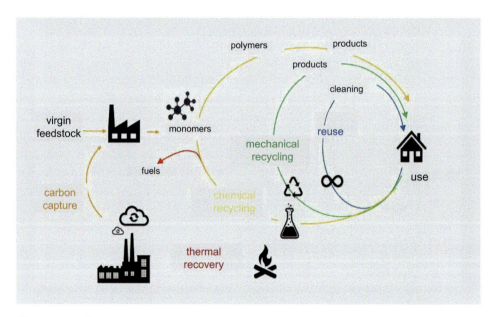

Figure 20.1 Scope for recycling and recirculating plastics

---

[1] Source: *https://www.bpf.co.uk/press/closed-and-open-loop-plastic-recycling.aspx*

20.2 Mechanical Recycling

Recycling of Thermoplastics

The recyclability of plastics depends on the type of plastic as well as the degree of contamination. Thermoplastics can be recycled by melting. They are therefore very well suited to mechanical recycling. If possible, the waste should be of one type of plastic so that good product properties can be achieved. Figure 20.2 shows the mechanical recycling process.

plastic type

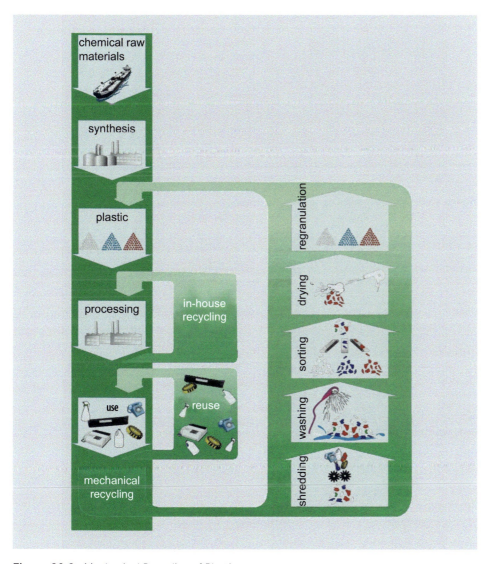

Figure 20.2 Mechanical Recycling of Plastics

in-house recycling	Starting from the production of plastics, there is the possibility of in-house recycling in the production of semi-finished plastic products or injection molded parts. Rejects and production waste, for example, are shredded and appropriately processed. They are then fed back into the direct production process.
multiple use	The utilization of plastic products for reuse, such as plastic deposit bottles or beer crates, Euro pallets, etc., is a second variant of mechanical recycling.
melting	During the remelting of waste plastic mixtures, certain plastics are destroyed by the required temperatures, while others do not melt at all. Figure 20.3 indicates the melting temperature ranges of PVC, PA and PC.

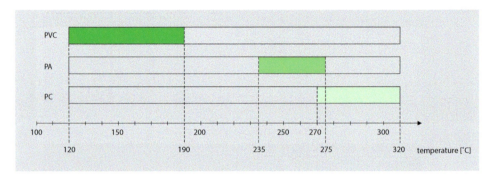

Figure 20.3 Melting temperature ranges of various plastic types

melting temperature range	PVC has a melt temperature range of 120 to 190 °C (248 to 374 °F), which in the case of PA is between 235 and 275 °C (455 and 527 °F). In the case of PC, from which the already mentioned CD is made, for example, it is between 270 and 320 °C (518 and 608 °F). This shows that it is not possible to find a common melting temperature for different types of plastic. For example, at a temperature of 250 °C (482 °F), the PVC has already decomposed, and the PC has not yet melted, while this temperature would be optimum for PA.
quality standards	Thus, it is not possible to produce a homogeneous plastic melt from this three-component mixture. The products made from it would therefore not be able to meet high quality standards.
contaminants	Contaminants adhering to the waste should be avoided or removed. Otherwise, they will be melted down along with the plastic waste as a foreign substance and thus lower the quality of the products. In the case of the yogurt pot, the percentage of contaminants by weight due to the remaining yogurt residue is often more than the weight of the container itself, which weighs only around six grams. So, when plastic waste is collected, it is often the case that more "contaminants" are colleted than the actual raw material "plastic," which then must be separated from the contaminants.

The best results in thermoplastics recycling are achieved when the waste to be recycled is completely unmixed, i.e., identical in terms of plastic grades, plastic types, additives, and fillers. Furthermore, the waste must be uncontaminated in order that high-quality products may be manufactured. Purity and compatibility of the starting materials are therefore of decisive importance in recycling:

grade purity

- *Pure type* means that only the plastic of one raw material manufacturer with the same type designation is used.
- *Pure grade* means plastics with the same specification according to DIN EN ISO 11469 or VDA 260.
- The term *similar grade* is used when the basic polymers (e.g., PP; PE) are the same, but the additives differ.
- *Blended* means that different plastics are processed which are chemically compatible (e.g., ABS and PC); in this case, homogeneous miscibility must be possible at least in the melt (see Figure 20.3).
- *Impure* means that the plastics to be reprocessed still contain substances from previous use that affect the properties of a new molded part.

The various existing processing technologies for plastic waste generally operate according to the basic process scheme shown in Figure 20.2: shredding, washing, sorting, drying, and repelletizing.

process diagram

Some examples of material recycling of technical parts after use are given here: recycled PET deposit bottles, bottle crates, dye tubes from the textile industry, control housings for heating systems, window profiles, rear and blinker lights, and bumpers of automotive vehicles.

examples

Recycling of Thermosets

The non-melting properties of thermoset materials prevent them from being recycled directly by remelting. The materials used consist of resin, hardener, fillers and reinforcing materials. The fillers and reinforcing materials account for considerable proportions, which can be up to 80% by weight of the total composition of the material. This fact is applied in so-called particle recycling for material recovery. They can thus be crushed (milled) and used as fillers in or together with thermosetting virgin material.

reinforcing materials

■ 20.3 Chemical Recycling

Figure 20.4 shows the "chemical" or "feedstock" recycling process. Here, plastic waste is shredded and processed under pressure and high temperature to produce gases, waxes, and oils, which in turn can be made available to the chemical industry.

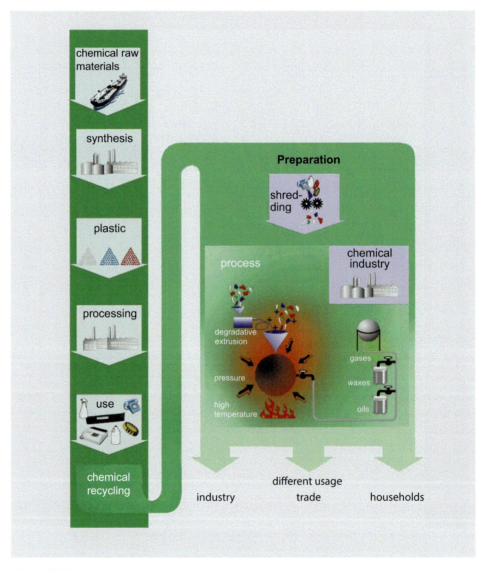

Figure 20.4 Chemical or Feedstock Recycling

Feedstock recycling of plastic waste can be achieved only within a very complex process. It plays merely a minor role in the context of material recycling of plastic waste, as its share in total waste was only 0.2% in 2019.

feedstock recycling

■ 20.4 Thermal Recovery

For heavily contaminated plastic waste, there will be no other recycling option than energy recovery, because the processing costs for heavily contaminated plastic waste would be so high that other recycling processes would not make sense either ecologically or economically.

energy recovery

Generally, energy recovery (Figure 20.5) is understood to mean thermal recycling. The thermal recycling of plastics aims at the utilization of the high energy content of the plastics, the reduction of the waste volume, the inertization of the solid incineration residues and the avoidance of an uncontrolled release of hazardous substances.

energy content

Dirty plastic waste thus serves as an important, so-called refuse-derived fuel (RDF). Without the high calorific value of plastics, which are made from petroleum, many incineration processes in waste incineration plants would not run stably. The Federal Environment Agency lists 68 waste incineration plants in Germany (as of 2016) that mainly burn municipal waste.

substitute fuels
secondary fuels
waste incineration plant

These substitute fuels are also referred to as secondary fuels. In addition to plastics, they also include paper, wood or treated sewage sludge. They are produced in a recycling process, reduce the use of fossil fuels, and are mainly used in industrial, heating and cement power plants. RDF can be solid or liquid and is obtained from waste mixtures from households, commerce, and industry. In Switzerland, there are 30 waste-to-energy plants (Kehrichtverwertungsanlage KVA), which already generate a significant amount of electricity and heat.

waste incineration in Switzerland

Of course, waste incineration plants also generate "waste" which is by no means unproblematic, and which must therefore be disposed of safely. Regarding air pollution control, which is affected by the combustion gases, waste incineration plants are governed by the Federal Pollution Control Act (BImSchG).

Federal Pollution Control Act (BImSchG), Germany

Energy recovery only makes sense for very heterogeneous waste mixtures (e.g., household waste) in which the plastics are just part of the mixture. Incineration of pure plastic waste is neither ecologically sensible nor technically possible in conventional incineration plants.

material recycling

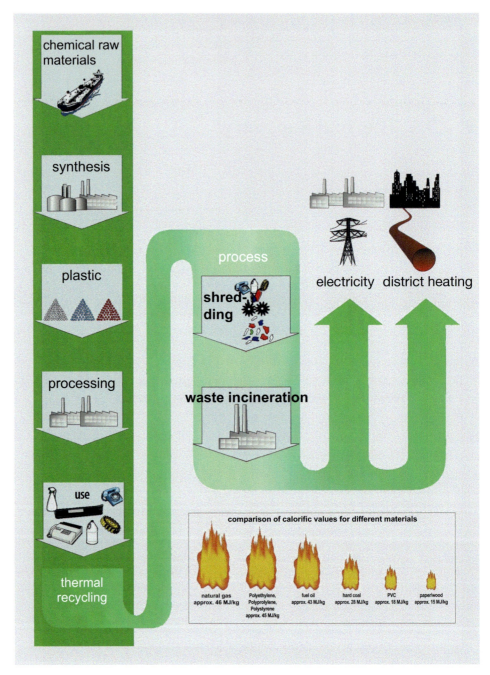

Figure 20.5 Energy recovery from plastic waste

20.5 Recycling of Plastic Waste

Figure 20.6 shows the development of plastic waste recycling in Europe from 1994 to 2019 in %. In this context, there is still a lot of potential for optimization towards mechanical and chemical recycling.

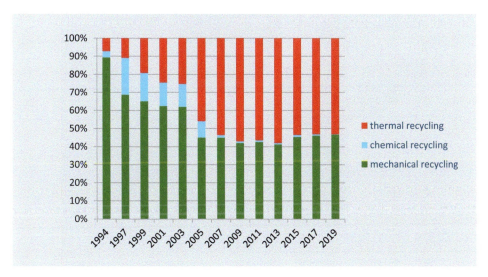

Figure 20.6 Development of plastic waste recycling in % (source: German Federal Environment Agency 2020)

The material flow verification for 2019 in Germany shows that a total of approximately 6.2 million tons of used sales packaging from households and small businesses was collected and sent for recycling. Some 99% of this annual plastic waste was recycled.

material flow

20.6 Examples and Products

Examples of the reuse and recycling of plastic waste are manifold. Recycled materials can be used to produce not only low-tech parts, but also high-quality products. Plastic recyclates are used across all industries and economic sectors in Germany. The construction sector dominates strongly, at 43% (Figure 20.7).

Other major consumers of recyclate are the packaging sector (24%) and agriculture (11%). The remaining lower percentage is accounted for by the automotive sector, furniture, and household goods.

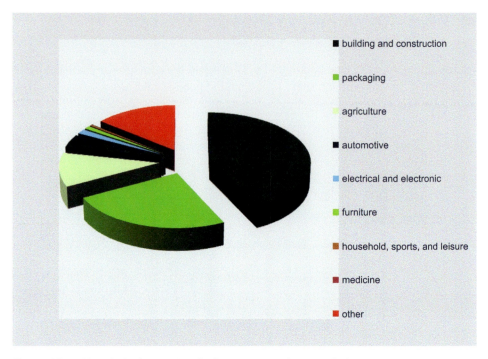

Figure 20.7 Use of plastic recyclate in Germany 2019 (source: Federal Environment Agency 2020)

Recycling of Plastic Window Profiles

window profiles (PVC)

One example of a successful circular product flow is that of window profiles, which are usually made of polyvinyl chloride (PVC). Burning PVC is unacceptable because of the toxic gases it produces, and landfilling is undesirable because of the lead stabilizer that is often used. The most environmentally friendly solution here is also recycling. The recycling scheme for old windows and window profiles made of PVC is shown in Figure 20.8.

lifetime

It was in the 1960s that the first plastic windows entered the economic cycle. As these are high-quality products with a service life of up to 40 years, they have only been produced as waste for about 20 years. Furthermore, they are available already sorted by type (only PVC) and usually only slightly soiled, or the frames can be easily freed from dirt. Therefore, their recycling rate, i.e., the proportion of the recyclable quantity of PVC windows, is now very high. Just like the window profiles, roller shutters made of PVC and doors can also be recycled. A window frame made of recycled PVC shows only minor differences in quality compared to a frame made of pure virgin PVC.

The growth in use of PVC recyclate from old plastic windows, plastic roller shutters and plastic doors is shown in Figure 20.9.

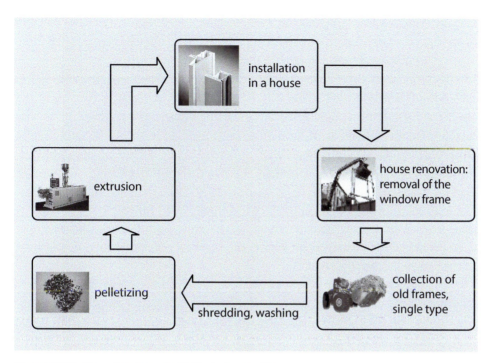

Figure 20.8 Recycling loop of a PVC window profile

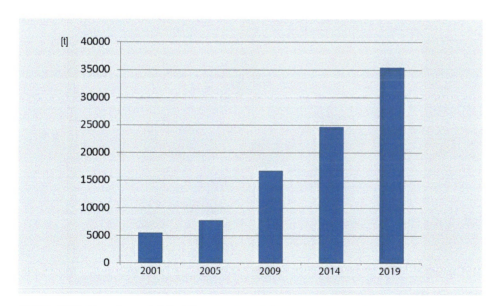

Figure 20.9 Use of recycled PVC from 2001 to 2019 (source: Rewindo)

The increase from 2001 (at nearly 5,500 t) to 2019 (slightly more than 35,000 t) is enormous. In 2016, almost 90% of the recyclable quantity of PVC windows, shut-

recycling of PVC

ters and bags made of PVC plastic was recycled. These figures strongly demonstrate that recycling is not only ecologically imperative, but also makes sense from an economic point of view.

Optical Storage Media Recycling (CD, DVD, Blu-ray)

CD/CD-ROM/DVD

The CD as such is a composite material consisting of three layers, the layer of clear PC that contains the music information, the reflective layer of aluminum and a lacquer layer to protect the CD.

CD/CD-ROM/DVD-Recycling

The recycling of optical storage media is now an established process. In principle, all CDs, CD-ROMs, and other data carriers made of polycarbonate (PC) can be recycled. Using a process developed by the Bayer company in the 1990s, the coating is removed from the polycarbonate without leaving any residue and is disposed of in an environmentally compatible manner. The pure PC regrind is then further processed into high-quality products.

Optical storage media such as CDs and DVDs are an example of high-quality recycling (Figure 20.10). The CD is a long-life product that only becomes waste after many years. Since CDs/DVDs are virtually pure grades (only one type of plastic) and not contaminated, they can be processed easily into new, high-quality plastic products.

Figure 20.10 CD/DVD and its packaging (Source: Netstal company)

CD covers

The situation is different for the three-part jewel cases of the CDs: the bottom and cover are made of clear PS; the insert that holds the CD in place is made of clear or colored PS. The CD's table of contents is made of paper. It is not glued to the plastic and can be removed. If the case parts are sorted according to color, it is possible to produce clear parts again from clear parts by shredding and remelting.

In addition to the highly tear-resistant films made of plastic used for packaging and selling optical storage media, there are of course other high-quality materials such as paper. This packaging must be recycled accordingly using other processes and put to appropriate use.

paper packaging

■ 20.7 Performance Review – Lesson 20

No.	Question	Answer Choices
20.1	Recycling of plastics is _____.	possible / not possible
20.2	In addition to mechanical and energy recycling, there is also _____ recycling.	ecological / chemical / biological
20.3	Mechanical recycling of waste can help to reduce the consumption of _____.	raw materials / energy and raw materials
20.4	Thermoplastics can _____ by melting them.	be recycled / not be recycled
20.5	Thermoset waste can be _____ and fed into the production process.	shredded / incinerated
20.6	Plastic waste is easier to recycle if it arises _____.	soiled / cleanly
20.7	Mechanical recycling of production waste into high-quality products is possible if the plastic waste is unpolluted and _____.	mixed / sorted by type
20.8	Pure grade recycled materials meet _____ quality requirements.	high / only low
20.9	Nowadays, optical storage media such as the CD/DVD are _____.	recyclable / not recyclable
20.10	Plastics should be considered as _____ secondary raw materials.	worthless / valuable
20.11	The different melting temperatures of mixed plastic waste play an _____ role in processing.	unimportant / important
20.12	Recyclable window profiles made of the material PVC are currently (values from 2017) recycled to _____ %.	50 / 70 / 90

21 Lesson
Qualification in Plastics Processing

Key Questions

Which professions and qualifications for plastics processing are available in industrial companies and skilled crafts?

What are the focus areas of the training?

What further training and promotion opportunities are there?

How is plastics training organized in the skilled crafts sector in Germany?

How can the qualification system for plastics processing in the skilled trades be organized?

Example Germany: What further training and career advancement opportunities are available for plastics processing professions?

Contents

21.1 Fundamentals

21.2 Plastics Training in Industry

21.3 Plastics Training in the Skilled Trades/Crafts Sector

■ 21.1 Fundamentals

In 2019, more than 1.6 million people worked in the plastics industry in approximately 60,000 companies in Europe. The employees generated sales of more than 350 billion euros. The plastics industry is made up of plastic manufacturers (chemical industry), plastics machinery manufacturers (mechanical engineering) and plastics processors (manufacturing industry).

plastics industry
plastic manufacturers
plastics machinery manufacturers
plastics processing companies

It would be very difficult to provide a compact coverage of the educational, training and qualification systems within the plastics sector for all of Europe, much more so for the entire world. However, as a representative case study, this chapter presents the educational structure and opportunities within plastics technology in Germany, which has the largest plastics industry within Europe, as well as those of its neighbors Austria and Switzerland.

The demand for qualified specialists continues steadily in the plastics processing industry, a continuously growing, forward-looking industry in the Federal Republic of Germany, with more than 310,000 employees, over €60 billion in sales and just under 3,000 companies in 2019. The industry, dominated by medium-sized companies, is characterized by high innovation and a diverse product range. However, solving the skills shortage, in both the skilled trades and in the plastics processing industry, is essential to continue the production of high-quality and competitive products and services, especially in a high-wage country such as Germany. This chapter presents the plastics occupations and qualifications and other aspects of skills development in the German plastics processing industry and in the skilled trades.

The manual skills aspect of plastics knowledge is explicitly dealt with here as it has a significant impact within a value chain on the overall quality of a product. For example, both the professional manufacture of plastic windows and their installation, as a service onsite by specialized skilled craftspeople, make a decisive contribution to the quality of the product "window", which after all has a service life of 40 to 50 years according to today's standard.

In Austria and Switzerland, the plastics processing industry also plays a significant role in the industrial sector. The number of companies in Austria is over 500 and the number of employees is around 29,000. In Switzerland, about 28,000 people worked in this industry in about 430 companies in 2019. In both countries, the structure of this industry is typically SMEs (small and medium-sized enterprises). Both countries have developed qualification systems for the plastics sector similar to those in the Federal Republic of Germany. These systems are therefore also presented in this lesson.

21.2 Plastics Training in Industry

The "dual vocational training" that is conducted in Germany is recognized worldwide. It is based on parallel training in the workplace and general and specialist vocational instruction at a vocational school. Trainees are employed and paid by the training enterprise.[1] Professions and professional qualifications result from the technical and interdisciplinary requirements that exist in the plastics processing industry, and are subject to constant change.

For plastics, the most important primary shaping processes used today are extrusion, calendering, coating of flexible carrier sheets, injection molding, compression molding, production of hollow bodies by various processes, reinforcement of reactive resins and foaming. *primary shaping processes for plastics*

For rubber, the main shaping processes are blending (compounding), extrusion, calendering, compression molding, spread coating, injection molding and conversion (e.g., in tire production). The products manufactured in this way generally must meet high quality requirements. *shaping processes for rubber*

Industrial plastic products are manufactured by non-cutting methods and usually in an interlinked process. If possible, their surfaces should not require any further finishing. Only by ensuring a smooth process flow can high product quality be guaranteed. *plastic products*

Economic aspects and, above all, environmental protection also require the work to be carried out in a way that conserves raw materials and energy. This is an increasingly important challenge for the plastics processing industry. The reuse (recycling) of scrap and waste is of particular importance here. *environmental protection*

The shaping of plastics always takes place under specific pressure, temperature and time values that must be precisely monitored. Only then can the production process run reproducibly and at the same time ensure high quality of the products. *quality*

Modern, technically advanced machines, for example in injection molding or extrusion companies, are employed in the plastics processing industry. The systems are equipped with mechanical, electrical, pneumatic and hydraulic control and regulation mechanisms. They are very cost-intensive and usually used in 24-hour shift operation. *machines*

Ultimately, the decisive factor for product quality is the forming tool, which is usually made of high-quality steel alloys. The qualified personnel must be familiar with the requirements imposed on the tool/mold in the production process and be able to assess them appropriately and take them into account accordingly. *tool*

In addition, there are the precise requirements that the "customer" imposes on the plastic product as well as the functions that are related to the item. *customer requirements*

This is where the overall value chain takes effect and the manufacturing industries, often skilled trade companies, come into play. For example, skilled trade companies install or replace the industrially produced plastic windows or produce the *value chain cooperation skilled trade*

---

[1] https://www.deutschland.de/en/topic/business/how-germanys-dual-vocational-training-system-works

windows themselves, especially in the case of tailor-made and one-off products for which industrial production is not worthwhile.

Professions in the Plastics Processing Industry

Two main professions have now established themselves in the plastics processing industry:

- Process technician for plastics and rubber engineering
- Materials tester with focus on plastics

Process Technician for Plastics and Rubber Engineering

For well over 20 years, the plastics processing industry in Germany has been developing a profession of key importance, the *process technician for plastics and rubber engineering*. It is an occupation with vocational training that has been officially recognized since 1997.

The duration of the training for this occupation is three years and it ends with a final examination, the "skilled worker examination" according to the *Dual System* of vocational education. The occupation is divided into different specializations (Table 21.1), corresponding to the specialization of the plastics processing industry. The specialization is specified on the final certificate.

Table 21.1 Fields of Specialization: Process Technician for Plastics and Rubber Engineering

Specialization	Examples
molded parts	toys, bottles
semi-finished products	tubes, films, sheets (which undergo further processing)
multilayer rubber parts	car tires, car door seals
compound/masterbatch production	production of polymer blends
construction components	technical components, e. g., chemical pumps
fiber composite technology	covers, sport articles, rotor blades for wind turbines
plastic windows	windows, shutters, doors

specializations in plastics industry

specialization: plastic windows

The specializations reflect different fields of specialization within the plastics processing industry. For example, thermoplastics are mostly produced by the plastics manufacturing industry (chemical sector) and supplied to the plastics processing industry as pellets (or, in the case of PVC, as powder). Rubber processors, by contrast, develop and produce their own "rubber compound" in-house. The high importance of windows made of plastic, which are in fact composed of plastic profiles with a steel reinforcement, is reflected in their own specialization.

Specialization in *semi-finished products* is particularly important for extrusion companies. However, the vocational training is so basic that a change to companies with other process specialties is usually possible at any time after a short training period.

specialization: semi-finished products

Process technicians for plastics and rubber engineering are first and foremost the specialist workers in the plastics processing industry, and that is where they are mainly employed.

plastics processing industry specialists

However, this qualification is also in demand outside the plastics processing industry in the so-called industrial fields of application. These include mainly the mechanical engineering, but also the automotive, medical, electrical and aircraft industries. These industries already use plastic parts in many applications and are increasingly turning to processing plastic molding compounds themselves.

industrial sectors

Depending on their qualifications and preferences, process technicians for plastics and rubber engineering work in the above-mentioned industrial sectors as machine setters, shift supervisors, machine operators, quality inspectors or in production monitoring.

jobs

A machine setter installs tools on machines, sets them up and controls the startup of the machines. Mold changing and the setting up of today's almost exclusively electronically regulated and controlled machines is very time-consuming and requires good knowledge of the machines and tools as well as a high sense of precision. The extrusion process is used in particular for mass products (plastic pipes, window profiles, packaging films, etc.). In mass production, tool changes are relatively rare. In the consumer goods sector, by contrast, the trend is toward a greater variety of products and smaller quantities. Accordingly, rapid conversion of machines will become increasingly important in this area in the future. Qualified set-up personnel are essential here.

machine setter

After setting up the machine and starting up production, the process technician for plastics and rubber engineering primarily assumes the task of machine guidance and thus the monitoring and control of the production machines and/or production systems, which today operate largely fully automatically for the extrusion sector. The technician becomes a "diagnostician" with a great deal of testing and monitoring responsibility. Malfunctions must be detected and localized at an early stage so that the necessary steps can be taken to remedy them to avoid major production breakdowns. He or she must be able to rectify minor malfunctions independently.

machine operator

As a shift supervisor, the process technician for plastics and rubber engineering independently manages complete work shifts and instructs new employees. This requires not only technical and scientific knowledge, but also a high level of social competence. This means that the supervisor must have the ability to work together (teamwork) and cooperate, assume responsibility, and instruct and lead employees.

shift supervisor

quality inspector	Another area of responsibility, which usually requires several years of professional experience, is quality assurance. Here, for example, a quality inspector carries out tests on flow behavior, heat tests and mechanical tests in the company's own laboratories, thus ensuring the perfect condition of the starting materials and the end products.
work planner	As an employee in the work planning department, he or she checks, among other things, the raw materials, and their preparation (mixtures) for production and ensures that the material is supplied to the machines.
area of responsibility production	Where the process technician for plastics and rubber engineering is employed in practice depends both on the chosen specialization and on the size and structure of the company of employment. In most cases, he or she now works on the production machine or production lines and is responsible for their operation and for their downstream equipment.
education basic vocational training specialization: semi-finished products	The separate specialization in *semi-finished products*, an example of which is presented here, shows the high importance of the extrusion process within the three-year training program. The first and second years of training consist of common basic education and general technical training. In the third year, training is then provided in the various areas of specialization. In the area of manufacturing semi-finished products by extrusion, the following learning objectives and contents are taught (Table 21.2).
learning objectives and contents	**Table 21.2** Summary of Learning Objectives and Learning Content: Extrusion

Learning objective	Learning content
perform system analysis of the extrusion line and the process	drives; plasticizing units and their temperature control systems; control and regulating equipment; automation techniques; production sequences; special processes; occupational health and safety
know the structure of extrusion dies	die types; temperature control systems; additional equipment
perform calculations for extrusion	throughput; speeds; cross sections
evaluate technical drawings	product drawings; overall, partial drawings; parts lists; technical documentation

extrusion materials example: plastic carrier bag and plastic window	Two particularly important sub-items in extrusion for the skilled worker are knowledge of the extrusion tools and of the extrusion process, because they decisively determine the quality of the extruded products.

- In this training area, all important plastics for the extrusion process are presented: both the inexpensive commodity plastics, such as polyethylene (PE) and polypropylene (PP), which are used in our example of the *plastic bag* or *plastic carrier bag*, and the high-performance material PVC, which is used in *plastic window profiles* and plastic roller shutters. The focus is on the processing of these materials and the processing parameters within the extrusion process.

- The process also requires knowledge of the effect of temperature control, the pressure in the various phases and the lubricants and release agents. It also entails mounting the dies and testing their function, as well as operating the extruder and the extrusion line. Another sub-item is the prevention of manufacturing defects, such as surface flaws, dimensional inaccuracies, or impermissible deformations.

processing parameters
extrusion die
die mounting

Materials Tester with Focus on Plastics

Ever-increasing customer and quality requirements and the diverse tasks in production have contributed to the emergence of a second nationally recognized plastics profession since 2013, the *materials tester specializing in plastics*. The vocational profile of the materials tester is offered in the following specializations:

materials tester with focus on plastics

- Focus: metal technology
- Focus: heat treatment technology
- Focus: systems engineering
- Focus: plastics technology

This profession is designed to help companies successfully meet the growing challenges of overall testing and monitoring.

The duration of the vocational training for this occupation is 3.5 years. The content of the training is as follows:

training content

- analysis of plastics properties
- performance of test procedures
- evaluation of test methods
- failure analysis
- structure and organization of the apprenticing company
- procurement, production, sales as well as administration of production materials and deliverables
- rights and duties of the trainees
- occupational health and safety
- environmental protection measures

Further Training and Promotion Opportunities

The process technician for plastics and rubber engineering works in an industrial sector that is constantly evolving. New materials, microelectronics and the trend toward automating entire production facilities are changing work processes, pro-

professional progress

duction methods and the organizational structures of companies. However, this also means that the demands on the qualifications of employees are changing. Control and monitoring activities, troubleshooting, but also communication and cooperation skills are becoming increasingly important. Simple, more manual skills, however, are becoming less important.

further training

At certain intervals, the process technician for plastics and rubber engineering should take part in continuous training courses and courses offered by various institutes of further education. This not only helps to keep employees up to date with the latest technology, but also increases their chances of advancement within the company.

Bachelor Professional of Plastics and Rubber Production

In addition, there is the possibility for him/her to be trained as a *plastics quality inspector* or as a *bachelor professional of plastic and rubber production* by means of further and advanced vocational training.

plastics technician

Attendance at a special technical school can lead to a degree as a *plastics technician*.

Bachelor of Engineering (B. Eng.)
Master of Engineering (M. Eng.)

Likewise, the *Bachelor of Engineering* (B. Eng.) or *Bachelor of Science* (B. Sc.) degree and the *Master of Engineering* (M. Eng.) or *Master of Science* (M. Sc.) degree at colleges and universities are fundamentally open to applicants, provided they have the necessary entry requirements (German *Abitur* (high-school diploma) *Fachabitur* (vocational diploma) or *Fachhochschulreife* (university of applied sciences entrance qualification), master craftspeople and comparably qualified individuals with vocational advanced further training). These new designations replace the former designations *Diplom-Ingenieur (FH)* at universities of applied sciences (*Fachhochschulen*) or *Diplom-Ingenieur* at technical universities and they are thus recognized throughout Europe (under the Bologna process).

Job Situation and Future Prospects

history

The profession of *process technician for plastics and rubber engineering* is the basic occupation in the whole plastics processing industry. It is a now well-established and has been a recognized vocational training occupation for over 20 years. Prior to 1976, there were three plastics occupations: *plastics mold makers, plastics fitter* and *rubber and plastics liner maker,* but these have been merged into the occupational profile of "process technician for plastics and rubber engineering" in the various specializations. However, there is still a considerable need for the process technician for plastics and rubber engineering, as the plastics industry continues to expand strongly.

processing technology

Plastics processing techniques are currently undergoing significant development, particularly with regard to the exact reproducibility of shaping and follow-up processes. Control and regulation techniques are being refined, processes are monitored by computers and increased attention is being paid to automation.

Knowledge and skills in "mechatronics", i.e., the combination of mechanics and electronics, are becoming increasingly important.

Trends in demand and industry development make the process technician for plastics and rubber engineering a profession that offers excellent prospects for the future, including for women, who, however, still account for only a small proportion of trainees to date. *(future prospects)*

This includes, in particular, additive manufacturing, which might possibly be included in the job description of the process technician for plastics and rubber engineering as a further specialization. *(additive manufacturing)*

There are other professions in the plastics processing industry and in the production of industrial plastics processing companies. The following are worth mentioning: *(other professions in the plastics processing industry)*

- Industrial mechanic – production engineering
- Industrial mechanic – maintenance
- Industrial electronics technician
- Mechatronics technician
- Industrial management assistant
- Tool engineer – molding

They all perform important duties for the overall product. However, the *process technician for plastics and rubber engineering* is the backbone in manufacturing because he or she must master the manufacturing process.

The requirements arising from the digital transformation and "Plastics Industry 4.0" are also having an impact on qualification requirements and qualification development throughout the plastics industry. In the future, networked thinking, and the inclusion of the entire system, including the customer, will play an increasing role. *(Industry 4.0)*

21.3 Plastics Training in the Skilled Trades/Crafts Sector

- **Vocational training**

    Plastics training in the skilled trades/crafts sector differs fundamentally from training in industry because the processing of plastics in the skilled crafts sector cannot be confined to one industry segment. Plastics are processed in almost all technically oriented skilled trades alongside conventional materials.

The special knowledge that is important for handling and processing the plastics in a way that is material-appropriate is taught in specially designed training courses.

- **Plastic training centers**

  At present, 43 so-called plastics training centers are located all over Germany, at which the plastics training courses developed by the Institute for Plastics Processing (IKV) in Aachen, Germany, are held (Figure 21.1).

Figure 21.1 Overview of the plastic training centers in Germany

- **IKV, Aachen**

    The Institute for Plastics Processing (IKV) in Industry and Craft at RWTH Aachen University, Seffenter Weg 201 in D-52074 Aachen, Germany, develops and supervises these plastics training courses, which are specially tailored to the skilled crafts and are currently run by the plastics training centers approved by IKV.

    The plastics qualification courses offered at these training centers can be accessed on the website under *https://www.ikv-aachen.de/en/industry-and-craft/course-programme-for-craft/*.

- **IKV Context System**

    With the so-called IKV Context System, which has proved itself in plastics qualification for over 50 years, a personal media network system is available that is tailored to the needs of adult learning in the field of skilled crafts.

    The learning elements cover the most important topics only, with missing-word exercises (cloze texts) and tasks motivating the course participants to concentrate on their work. In addition to being an effective teaching aid, which facilitates the learning process for the course participants in particular, the self-completed context provides them with a reference work with which they are familiar, and which also contains data on the processing methods that they can refer to directly if necessary. Figure 21.2 shows an example of such a context page.

More than 20,000 people are trained each year in the Federal Republic of Germany using this context system and thereby become qualified for the special requirements in skilled crafts.

quantity of participants in training courses

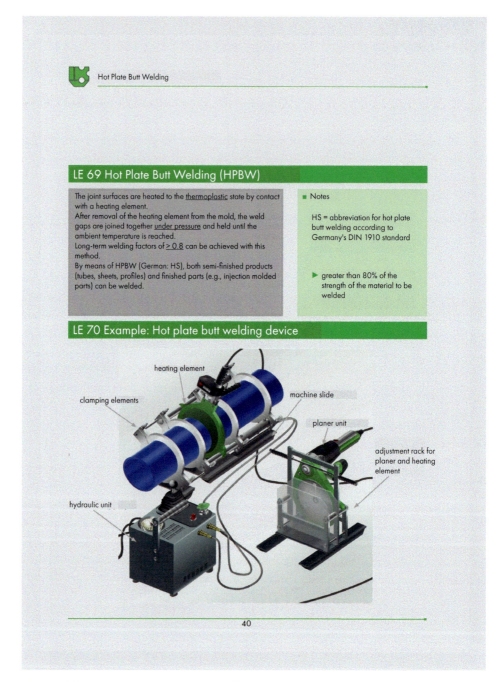

Figure 21.2 Example of a context page with cloze text

Appendix
Selected Literature

AKI (Ed.)	Kunststoffe – Werkstoffe unserer Zeit. Plastics Europe, Frankfurt/Main o.J. (18th edition)
Abts, G.	Einführung in die Kautschuktechnologie. Carl Hanser Verlag, Munich 2019 (2nd edition)
Abts, G.	Kunststoffwissen für Einsteiger. Carl Hanser Verlag, Munich 2020 (4th edition)
Baur, E./Osswald, T./ Rudolf, N. (Eds.)	Plastics Handbook – The Resource for Plastics Engineers. Carl Hanser Verlag, Munich 2019 (5th edition)
Baur, E./Harsch, G./ Moneke, M.	Werkstoff-Führer Kunststoffe. Carl Hanser Verlag, Munich 2020 (11th, updated edition)
Bonnet, M.	Kunststofftechnik. Springer Vieweg Verlag, Berlin, Heidelberg 2016 (3rd edition)
Bonten, C.	Plastics Technology – Introduction and Fundamentals. Carl Hanser Verlag, Munich 2019
Braun, D.	Kleine Geschichte der Kunststoffe. Carl Hanser Verlag, Munich 2017 (2nd, updated and expanded edition)
Bruder, U.	User's Guide to Plastic. Carl Hanser Verlag, Munich 2019 (2nd edition)
Dahlmann, R./Haberstroh, E./ Menges, G.	Werkstoffkunde Kunststoffe. Carl Hanser Verlag, Munich 2021 (7th, fully revised edition)
Domininghaus, H.	Kunststoffe – Eigenschaften und Anwendungen. Springer Verlag, Berlin-Heidelberg-New York 2012 (8th, new and revised edition) (Elsner, P./ Eyerer, P./Hirth, T., Eds.)

Ehrenstein, G. W.	Polymer-Werkstoffe. Carl Hanser Verlag, Munich 2011 (3rd, fully revised edition)
Ehrenstein, G. W.	Polymeric Materials – Structure, Properties, Applications. Carl Hanser Verlag, Munich 2001 (older English edition of the above)
Ehrenstein, G. W.	Faserverbund-Kunststoffe. Carl Hanser Verlag, Munich 2006 (2nd, fully revised edition)
Endres, H.-J./ Siebert-Raths, A.	Engineering Biopolymers – Markets, Manufacturing, Properties and Applications. Carl Hanser Verlag, Munich 2011
Erhard, G.	Designing with Plastics. Carl Hanser Verlag, Munich 2006
Franck, A.	Kunststoff-Kompendium. Vogel Verlag, Würzburg 2011 (7th edition)
Gebhardt, A./Hötter, J.-S.	Additive Manufacturing. Carl Hanser Verlag, Munich 2016
Gebhardt, A./Kessler, J./ Thurn, L.	3D Printing – Understanding Additive Manufacturing. Carl Hanser Verlag, Munich 2018 (2nd edition)
Greif, H./Limper, A./ Fattmann, G.	Technologie der Extrusion. Carl Hanser Verlag, Munich 2018 (2nd edition)
Hellerich, W./Harsch, G./ Baur, E.	Werkstoff-Führer Kunststoffe. Carl Hanser Verlag, Munich 2019 (11th edition)
Hopmann, C./Michaeli, W.	Einführung in die Kunststoffverarbeitung. Carl Hanser Verlag, Munich 2017 (8th, revised and expanded edition)
Hopmann, C./Michaeli, W./ Greif, H./Ehrig, F.	Technologie des Spritzgießens. Carl Hanser Verlag, Munich 2017 (4th edition)
Michaeli, W./Greif, H./ Kretzschmar, G./Ehrig, F.	Training in Injection Molding. Carl Hanser Verlag, Munich 2001 (2nd edition; older English edition of the above)
Hüning, A./Schulze, M.	Die neue EG-Maschinenrichtlinie. (2006/42/EG). DC Verlag e. K., Bochum 2015 (3rd, revised edition) (Sichere Maschinen in Europa Bd. 5)
Kaiser, W.	Kunststoffchemie für Ingenieure. Carl Hanser Verlag, Munich 2023 (6th edition)

Köster, L./Perz, H./ Tsiwikis, G.	Praxis der Kautschukextrusion. Carl Hanser Verlag, Munich 2007
Lehder, G.	Taschenbuch Arbeitssicherheit. Erich Schmidt Verlag, Bielefeld 2011 (12th, newly revised edition) (established by R. Skiba)
Limper, A. (Ed.)	Verfahrenstechnik der Themoplastextrusion. Carl Hanser Verlag, Munich 2012
Lindner, C.	Produktion, Verarbeitung und Verwertung von Kunststoffen in Deutschland 2015. Alzenau 2016 (publ. by BKV GmbH; PlasticsEurope Deutschland e. V.; IK Industrievereingung Kunststoffverpackungen e. V.; VDMA e. V./Fachverband Kunststoff- und Gumminmaschinen; bvse-Bundesverband Sekundärrohstoffe und Entsorgung e. V.)
Osswald, T./Menges, G.	Materials Science of Polymers for Engineers. Carl Hanser Verlag, Munich 2012 (3rd edition)
Richter, F.	Regularien in der Kunststoffindustrie. Vogel Verlag, Würzburg 2012
Röthemeyer, F./Sommer, F.	Kautschuktechnologie. Carl Hanser Verlag, Munich 2013 (3rd, newly revised and expanded edition)
Rudolph, N./Kiesel, R./ Aumnate, C.	Understanding Plastics Recycling – Economic, Ecological, and Technical Aspects of Plastic Waste Handling. Carl Hanser Verlag, Munich 2020 (2nd edition)
Schwarz, O./Ebeling, F. W./ Richter, F.	Kunststoffkunde. Vogel Verlag, Würzburg 2015 (10th edition)
Schwarz, O./Ebeling, F. W.,/ Furth, B.	Kunststoffverarbeitung. Vogel Verlag, Würzburg 2009 (11th, revised edition)
Seul, T./Roth, S.	Kunststoffe in der Medizintechnik. Carl Hanser Verlag, Munich 2020
Thielen, M.	Extrusion Blow Molding. Carl Hanser Verlag, Munich 2021
Thielen, M./Gust, P./ Hartwig, K.	Blasformen von Kunststoffhohlkörpern. Carl Hanser Verlag, Munich 2019 (2nd, updated edition)

Miscellaneous Sources and Information

BKV GmbH	Bundesverband Kunststoff Konzepte Verwertung GmbH, Frankfurt/Main
	www.bkv-gmbh.de
bvse e. V.	Bundesverband Sekundärrohstoffe und Entsorgung e. V., Bonn
	www.bvse.de
CAMPUS database	CAMPUS (Computer Aided Material Preselection by Uniform Standards) is a registered trademark of Chemie Wirtschaftsförderungsgesellschaft mbH (CWFG), Frankfurt/Main
	www.CAMPUSplastics.com
DGUV 100-001	Accident prevention regulation
	German social accident insurance
	Principles of Prevention
DIN 65148	Aerospace; testing of fibre-reinforced plastics; determination of interlaminar shear strength by tenside test
DIN EN ISO 9001	Quality management systems – Requirements (ISO 9001:2015); German and English version EN ISO 9001:2015
DIN EN ISO 10350-1	Plastics – Acquisition and presentation of comparable single-point data – Part 1: Moulding materials (ISO 10350-1:2017)
DIN EN ISO 11403-1	Plastics – Acquisition and presentation of comparable multipoint data – Part 1: Mechanical properties (ISO 11403-1:2021)
DIN EN ISO 11469	Plastics – Generic identification and marking of plastics products (ISO 11469:2016)
DIN EN ISO 14001	Environmental management systems – Requirements with guidance for use (ISO 14001:2015)
	Austria: ÖNORM EN ISO 14001 (2009)
	Switzerland: SN EN ISO 14001:2005 (2005)
FICO	Fachverband der Chemischen Industrie Österreichs (FICO), Vienna
	www.kunststoffe.fcio.at

GKV e. V.	Gesamtverband Kunststoffverarbeitende Industrie e. V. (GKV), Berlin
	www.gkv.de
IK e. V.	Industrievereinigung Kunststoffverpackungen e. V. (IK), Bad Homburg
	www.kunststoffverpackungen.de
IKV	Institut für Kunststoffverarbeitung (IKV) in Industrie und Handwerk an der RWTH Aachen
	(Institute for Plastics Processing in Industry and Craft at RWTH Aachen University)
	www.ikv-aachen.de
KUNSTSTOFF.swiss	KUNSTSTOFF.swiss (formerly Swiss Plastics), Aarau
	KUNSTSTOFF.swiss
Kunststoffverarbeiter WKO	Bundesinnung der Kunststoffverarbeiter, Wirtschaftskammer Österreich (WKO), Vienna
	www.kunststoffverarbeiter.at
Official Journal of the European Union	DIRECTIVE 2006/42/EC OF THE EUROPEAN PARLIAMENT AND OF THE COUNCIL of 17 May 2006 on machinery, and amending Directive 95/16/EC (recast)
	https://eur-lex.europa.eu/legal-content/DE/TXT/?uri=CELEX%3A32006L0042
PlasticsEurope	Association of European Plastics Manufacturing Sector
	www.plasticseurope.org
Rewindo GmbH	Rewindo. Fenster-Recycling-Service, Bonn
	www.rewindo.de
UBA	Umweltbundesamt, Dessau-Roßlau
	www.umweltbundesamt.de
VDMA Kunststoff- und Gummimaschinen e. V.	Verband Deutscher Maschinen- und Anlagenbau e. V., Fachverband Kunststoff- und Gummimaschinen, Frankfurt/Main
	www.vdma.org
Wikipedia	The free encyclopedia
	www.wikipedia.org

Appendix
Glossary of Plastics Technology

3D printing	In 3D printing (3DP), special printers produce a workpiece layer by layer. The basis for this is a so-called plastic filament. Plastic is therefore a typical material for this manufacturing process. (See also: additive manufacturing.)
additive manufacturing	The term additive manufacturing (AM), for which the terms generative manufacturing or rapid manufacturing (RM) are also used, is a broad term for all manufacturing processes in which the material is applied layer by layer. In addition to plastics, metals are also used in these so-called "rapid technologies".
aluminum	Aluminum is a chemical element with the element symbol Al and atomic number 13. It is a silvery-white light metal and therefore a competitor product for lightweight plastics.
amorphous	Amorphous means without (regular) shape, glassy, not crystalline, a state of highest disorder, or structureless.
anisotropy	Directional structure (of the smallest particles), the consequence of which are directional properties (i.e., properties are different in different directions).
axial	Axial means in the direction of the axis.
biopolymers	These are polymers that are not based on petroleum (crude oil), but are produced based on cellulose, starch, etc., as well as on plant and animal proteins.

bonds	Bonding is a term used in chemistry. The chemical bond is a physico-chemical phenomenon by which two or more atoms or ions are firmly bound together to form chemical compounds.
calender (calendering)	The calender is a machine for the continuous production of thermoplastic, continuous, semi-finished products (film), sheets, floor coverings, etc., also for laminating.
calorific value	The calorific value is defined as the quantity of heat resulting from the combustion of 1 kg of solid or liquid fuel or 1 m^3 of gaseous fuel.
CAMPUS	The acronym CAMPUS stands for Computer Aided Material Preselection by Uniform Standards and is a computer-aided database for the properties of plastics. This database is regarded as the world leader in terms of standardization and thus the comparability of characteristic values and characteristic value diagrams. The database is based on ISO 10350 for single-point properties (e.g., density) and ISO 11403 for diagrams (e.g., stress-strain diagram).
carbon fiber reinforced plastics	Carbon fiber reinforced plastics (CFRP) are composite materials with carbon fibers as reinforcing material and a polymer matrix.
catalysis	Catalysis means the acceleration of a chemical reaction by catalysts (Greek: katálysis = dissolution).
cavity	Cavity is the term used to describe the specially shaped open space in a mold into which the material is filled or injected under pressure.
cellulose	Cellulose is the most commonly occurring carbohydrate. Cotton, jute, flax, and hemp are almost pure cellulose.
centric	Located in the middle, centered.
chain units	A chain unit is the repeating unit in a macromolecule.
chemical bond	Cohesive forces between atoms in molecules due to the drive to achieve an energetically favorable state.

Circular Economy Act	The Circular Economy Act (German abbreviation: KrWG) regulates, among other things, the recycling of plastic waste. It is the central Federal law of German waste legislation and regulates the dual system (waste system) in Germany.
compounding	Compounding refers to all operations to which a raw material is subjected before the molding compound is fed to the actual processing to produce molded parts.
compression zone	Designates the section of the screw in which the molding compound is compacted; it is therefore also called the compaction zone.
continuous	System variables change continuously, i.e., in randomly short-time intervals. Continuous is the opposite of discontinuous, such as discontinuous tube extrusion with a melt accumulator die.
Control unit	The control unit coordinates the movements of a machine and is usually housed in a separate control cabinet next to the machine.
convection	Convection, or heat flow, is one of the three mechanisms for transferring heat from one place to another, along with conduction and radiation (see there). Convection is caused by a flow that is carrying particles.
cooling section	Section in which the hot extrudate is cooled by a water bath, spray bath or on a chill roll to obtain the necessary degree of stabilization.
cross-linking	This is the linking of polymer molecules (macromolecules) by primary valences to form a usually three-dimensional network. Crosslinking of suitable plastics can also be achieved chemically by the addition of appropriate bridge-building polymers.
crystal	"Kristallos" (Greek) means ice or quartz. Crystals are solids having periodically arranged building blocks (atoms or molecules), bounded by flat surfaces. A state of ultra-high order.

crystalline	Made up of numerous very small imperfectly formed crystals (crystallites).
crystalline melting temperature range	In the crystalline melting temperature range (CMT), the crystalline regions of a semi-crystalline thermoplastic undergo melting. The abbreviation T_c is also used for this in the literature.
cylinder	The cylinder or barrel of an extruder accommodates the rotating screw. The screw and barrel together form the plasticizing unit of an extruder.
decomposition temperature	The decomposition temperature range (DT or also T_d) refers to the temperature range above which a material is destroyed by chemical decomposition.
degassing	Removal of volatile components from plastics, especially in the molten state. Degassing is used in compounding as well as in processing.
degree of automation	A measure of the conversion of a manufacturing plant to fully automated production.
delamination	Detachment of the fiber from the matrix or a matrix crack parallel to a laminate layer.
dispersion	Particles finely dispersed in suspension in another substance.
dissipation process	Dissipation processes are processes in which friction is converted into heat.
distillation	Main separation process in chemical technology, whereby liquid or liquefied substances are separated from others by evaporation and recondensation.
dosing	Dosing means measuring, metering, adding specific quantities.

Glossary of Plastics Technology

dual vocational training
(also: dual system of vocational education)

Dual vocational training is a system of vocational qualification in the Federal Republic of Germany. It is characterized by the dual learning locations of company and vocational school. The person to be qualified is called an apprentice (German: "Azubi"). The prerequisite for vocational training in this system is a training contract in Germany and an apprenticeship contract with an approved training company in Austria, Switzerland, and South Tyrol (Italy). The practical part of the training takes place in the company and the theoretical part in the vocational school. The two places of learning guarantee a high level of training, since not all companies in industry and skilled trades (micro-enterprises) have sufficient company infrastructures for comprehensive training in accordance with the criteria of the Vocational Training Act (German: BBiG).

Duales System Deutschland
(DSD, Dual System Germany)

Based on the German Packaging Act (VerpackG), and since 2019 on the Packaging Ordinance (VerpackV), the Dual System Germany (DSD) operates the disposal and recycling of packaging waste (mostly in yellow containers) in addition to municipal waste collection. Waste disposal in Germany is thus dual, i.e., divided into two segments.

elastomer

Spatially poorly crosslinked plastic, non-meltable, insoluble, but swellable. Designation of a class of plastics.

exothermic reaction

A chemical reaction in which heat is released.

extruder

The core of an extrusion line, which can have different designs depending on the application. The material enters through a radial barrel opening (feed hopper). The extruder has a stationary heated cylinder in which a rotating screw compresses, melts, homogenizes the added material and then presses it through a die.

extrusion	Extrusion is one of the most important processes in plastics processing. It is used for the continuous production of semi-finished products (pipes, sheets, profiles, and films). Extrusion is a primary forming process.
factory 4.0	The term "Factory 4.0" is a relatively new and sometimes controversial concept. It characterizes the current developments of advanced industrial societies, according to which machines and people cooperate more and more in a networked system. This includes the "Internet of Things" as well as digital factory models that create only a virtual image of the real world. Machines and plants become quasi "intelligent" and thus make "their own" decisions.
feed zone	First part of a screw in which the raw material is drawn in and transported into the next screw section.
fiber-reinforced composite	Fiber-reinforced composite (FRC) is a generic term for plastics in which fibers (aramid fibers, glass fibers, carbon fibers) are incorporated in the plastic, mostly in thermosets (duromers), but also in thermoplastics.
filament	A filament is a continuous thread of a certain diameter. In the case of natural fibers, silk is an example, and in the case of synthetic fibers, artificial silk. The opposite of this is a fiber of finite length. Examples are wool or cotton for natural fibers and cell wool for synthetic fibers.
flow curve	Represents the relationship between shear stress and shear rate of a fluid.
flow line	Occurs when partial streams of the melt flow together. It can form a weak point (strength reduction) in the semi-finished product.
flow temperature	Above the flow temperature (FT, but also T_f) the thermoplastic can be formed by applying only small forces.

fluid injection technology

Fluid injection technology (FIT) is the generic term for techniques used to manufacture components with larger wall thicknesses as well as components with hollow spaces. The fluid used can be either gas in gas injection molding (GIT) or water in water injection molding (WIT).

fluidized bed process

A dust-like or fine-grained material (e.g., silica sand) can be agitated by gases rising up through it at a certain characteristic flow velocity in such a way that the system resembles a liquid in many of its properties. In the case of plastics pyrolysis, the process enables rapid heat transfer and the process to be carried out in closed reactors.

functional groups

Groups of atoms that give chemical compounds a certain character and enable their classification into classes of substances having corresponding chemical properties (e.g., hydroxyl groups of alcohols, carboxyl groups of organic acids, amino groups of amines).

gas-assisted injection molding

Gas injection molding (GIT) is a special injection molding process in which a hot molding compound is pre-injected, and, in a second step, an inert gas (very slow-reacting gas) is used to press the melt into the rest of the mold. This process is used to save material for large-volume or low-cost injection molded parts.

gel coat

This is the resin layer, usually colored, which protects the underlying resin/fiberglass laminate from external influences, e.g., impact, UV light, chemicals, etc. After demolding, the gel coat is the visible or outer surface of the molded part. Therefore, it is applied as the first layer to the mold.

German Packaging Act	The German Packaging Act (VerpackG) of 2019 is a Federal regulation that aims to prevent or reduce the impact of waste from packaging on the environment. This regulation is based on the European Packaging Directive 94/62/EC (abbreviated PACK) regulating the placing of packaging on the market. The relatively new law replaces the German Packaging Ordinance (VerpackV).
glass fiber reinforced plastics	Glass fiber reinforced plastics (GRP) are composite plastics in which glass fibers are embedded in a polymer matrix as reinforcing material.
glass transition temperature range	In the glass transition temperature range (T_G), the amorphous areas of a thermoplastic soften when heated. Today, the abbreviation ST (softening temperature range) is often used instead (see there).
Grüner Punkt (Green Dot)	A so-called "Green Dot" is used to mark packaging waste that can be recycled and is collected within the framework of the Dual System Germany (see there).
hardener	A hardener (curing agent) is the second chemical component required to initiate the cross-linking reaction of the prepolymers to produce thermosets or elastomers.
holding pressure	The holding pressure conveys melt when the molded part solidifies during the injection molding process. This reduces volume shrinkage when the injection molded part cools down and compresses the microstructure.
homogeneity	Similarity, uniformity of composition throughout.
hydraulic	Hydraulic means working with the pressure of liquids (Greek: "hydro" = water).
injection molding	A molding procedure whereby heat-softened plastic material is forced or injected through a nozzle into a mold. The mold is cooled for thermoplastic material or heated for thermoset or rubber material.

injection molding cycle	The injection molding cycle is the total time of all operations performed within the injection molding machine that are necessary to manufacture a part.
injection pressure	This is the pressure applied to the molding compound by the screw during the injection process into the mold.
integrated management system	An integrated management system combines several individual management systems (quality, environment, safety, information protection …) into a comprehensive management system for the entire company.
investment	Long-term investment of capital to replace used production equipment (replacement investment) and to acquire new production equipment (new investment).
isotropy	Isotropic properties of substances are completely independent of direction (isotropic); they are the same in all directions.
just in sequence (JIS)	JIS is the most consistent form of just-in-time (JIT) delivery to a company's own production operations, usually within the assembly lines of automotive companies. There, complete assemblies such as seats are delivered to the assembly line in the order (sequence) in which they are to be installed.
Just in Time (JIT)	Just in Time (JIT) means that parts manufactured by suppliers are delivered to the assembly line on time (e.g., in the automotive industry) so that production does not stop, but also no (expensive) intermediate storage of the supplied parts is necessary.
laminar	Refers to a flow that occurs in layers (laminar = layered) without turbulence.
laminate	Designates the cured thermoset matrix or the cooled fiber reinforced composite (with a thermoplastic matrix).

laminating	Laminating is the joining of identical or different layers or films, possibly with the use of adhesives. However, this also includes the application of cover layers (fusible layers) to sheets, films, or fabrics.
land	This is the bearing surface along the top of the flights of a screw.
layer structure	Refers to the structure and arrangement of the individual layers (Latin: lamina) of a fiber-reinforced plastic (FRP).
locking force	The locking force is the force required to close the mold during the filling phase in the case of thermoplastics or during the curing phase in the case of thermosets.
logarithmic	Instead of linear values (the distances between the values are equal) on a scale (e.g., on the x-axis), logarithmic scaling of axes uses powers (10^1, 10^2, 10^3 etc.).
macromolecule	A macromolecule is consisting of threadlike or three-dimensional giant molecules with at least 1,000 atoms. These also include several natural substances such as cellulose, proteins and rubber.
masterbatch	Masterbatch is a colorant concentrate in a solid (plastic) or also liquid medium, which is added to the natural-colored granules during processing.
matrix material	The matrix material is the material that binds the fibers (glass, carbon, aramid).
melt	Molten molding compound.
melt flow index	The melt flow index is used to measure the viscosity of a plastic melt. Today, the term melt flow rate (MFR) or melt volume rate (MVR) is used. Thermoplastic polymer is forced through a 0.0825-inch orifice for 10 minutes according to ASTM D 1238.

melt volume rate	The melt volume flow rate (MVR) is used to characterize the viscosity or flow behavior of a thermoplastic. Special test equipment is used to determine the MVR (see also: MFR).
metering zone	Last zone of a screw, which is intended to homogenize the material again and ensure uniform temperature conditions. This zone gives the melt the pressure necessary to overcome the subsequent resistances and determines the output rate.
modulus of elasticity	The modulus of elasticity (E modulus) is the constant ratio of stress to strain in the elastic range of a material. It can be determined in the tensile test, compression test and flexural test. Because of the viscoelastic behavior of plastics, the time dependence must be considered for plastics.
molded part	The molded part is the plastic part produced by primary forming, which can often be used without any rework.
molding compound	The molding compound is an unshaped or preformed material that can, within certain temperature ranges, by means of noncutting methods of shaping (primary shaping), be processed into molded materials. This may be a molded part or a semi-finished product.
molecular weight distribution	The macromolecules of a plastic are of different lengths. From the molecular weight distribution, one can read how often a macromolecule of a certain length (of a certain weight) occurs.
molecule	A molecule is the smallest unit of a chemical compound. Molecules are composed of atoms, the smallest particles of the elements, which cannot be divided chemically. Molecules consist of two or more atoms bonded together. Molecules can be broken down again into their constituent parts by chemical methods.

monomer	The monomer is the basic building block (Greek: individual part) of which macromolecules are composed. For example, ethylene is the monomer of polyethylene.
multifunctional	Several functions are combined in one component. For example, the cable of a ceiling lamp carries the lamp body and conducts the electrical energy; it thus performs two different functions.
newtonian fluid	The viscosity of this particular fluid is constant and independent of the shear rate (e.g., water).
nonwoven	Nonwoven according to German standard DIN 61850 is a non-woven, solid sheet made of bundled glass filaments or glass staple fibers (surfacing mat).
number of flights	If several lands run around the core, the same number of flights is obtained. The number of screw channels is indicated by the number of flights.
orientation	In polymers, this refers to the occurrence of one or more preferred directions of the monomeric building blocks of the chain molecules.
orthotropy	This is also known under the term orthogonal or rhombic anisotropy. The properties are direction dependent. There is a symmetry of the properties to a system of three perpendicular (orthogonal) planes.
pellet	Pellet is defined as a plastic starting material in pellet form, usually in cylindrical or lens shape form.
petrochemistry	Collective term for large-scale industrial, chemical, or physico-chemical conversions of crude oil (petroleum) as a raw material base.
plasticize	Plasticizing is the process of converting plastics into a thermoplastic state by applying heat. The heat can be generated by external heaters and/or inner friction.

plasticizers	Plasticizers are substances that have a softening effect. In physical terms, plasticizing means decreasing the glass transition temperature range (T_G or also ST) of high-polymer materials to lower values, generally below room temperature. This transforms a rigid, brittle material into a flexible, rubber-elastic material (at room temperature).
plasticizing unit	The screw and barrel together form the plasticizing unit.
polarity (of plastics)	In chemistry, polarity refers to the formation of separate centers of charge within groups of atoms due to charge displacements. As a result, atom groups are no longer electrically neutral and thus become polar.
polyaddition	A chemical reaction in which the reactive groups or ends of monomers react with each other by migration of H atoms to form polymers (mnemonic: change of place).
polycondensation	Like polyaddition, except that water or another low molecular weight substance is split off during the reaction. However, migration of atoms or groups of atoms does not take place.
polymerization	Refers to a chemical reaction in which polymers are formed from monomers through the dissolution of the double bonds (C=C).
polymers	Long molecular chains are formed from monomers. The monomers are found as a recurring unit in the chain (Greek: "poly" "meros" = many parts).
polyolefins	Refers to a group of plastics that are made up of the elements C or H only. Examples are polyethylene (PE) and polypropylene (PP), which are among the most widely used standard plastics.

prepregs	According to German standard DIN 61850, prepregs are molding compounds made from flat or linear textile glass reinforcing materials. They are pre-impregnated with curable resin compounds. The molding compounds prepared in this way are usually glass fiber mats or glass filament fabrics, which are processed into molded parts or semi-finished products by hot press molding.
primary molding process	Primary molding (also: primary shaping) processes include injection molding and extrusion, for example, as the raw material is melted and given a new shape.
process	A process is a sequence of associated work steps that change the state of an "object". Each process consists of an "input", a transformation, and an "output". The goal is the value creation of an object.
process technician	Designation for a group of technical-commercial occupations. In the plastics sector, the job title is *process technician for plastic and rubber engineering* (formerly *plastics molder* and *plastics fitter* as well as *rubber and plastics liner maker*). In 2013, a second occupation in plastics processing technology was added: *materials tester/focus: plastics processing*.
pyrolysis	This is the term used for the thermal decomposition of chemical compounds.
quality assurance	Quality assurance (QA) stands for the specific technical and organizational measures (e.g., random sampling) to create and maintain the defined quality of the products or service within the quality management system.
quality management	Quality management (QM) refers to all organizational measures that serve to improve process quality, services and thus products of any kind. Since the focus is on the processes (procedures in the company), it is now also referred to as PQM (process-oriented quality management).

quality management system	Quality management system (QMS) comprises the planning, steering, control, improvement, and assurance of the quality of products and services. A well-known example of QM models is, for example, the process-oriented DIN EN ISO 9001.
quasi-isotropic	Almost identical properties in all directions. For fiber reinforced composites, this can be achieved by at least three directions of reinforcement with similar layer thicknesses.
raw material	Raw material is a naturally occurring starting material (e.g., coal, ores, wood, animal skins, cotton, but also water and air) for a skilled trade or industrial product. During the manufacturing process, the intermediate product (semi-finished product) is created, from which the finished goods (manufactured products) are made.
reaction-injection-molding	Reaction injection molding (RIM) refers to an integrated mixing and injection process for plastics with two or more highly reactive components.
recyclate	The term recyclate is a generic term for processable plastics with defined properties.
recycling	This is the reutilization of raw or used materials. For example, sorted waste is recycled by processing it into pellets and feeding it back into the extrusion process or injection molding process in proportions.
recycling rate	The recycling rate indicates the proportion of a material or waste material (plastic, steel, aluminum, paper, etc.) that is recycled.
refining	Purification and upgrading of natural substances and technical products (sugar, petroleum or similar). Refining takes place in a refinery.
Regranulated material	Regranulated material is obtained via a melting process as granules of uniform grain size and no dust content.

residence time This is the time that a particle of a plastic melt is subjected to heated machine parts (e.g., extruder, mold). Since not all particles flow through the same path and at different velocities, there is always a residence time distribution.

resin Resin is an amorphous material with a soft to rigid consistency. Thermosetting resins form the basis for thermoset plastics.

resin transfer molding This process is used to produce molded parts from resin (RTM process). Reinforcing materials (glass fibers, carbon fibers) are first placed in the molds. The mold is then closed, heated, placed under vacuum and finally the resin is injected into the mold. After curing, the part can be removed.

resistance heating An electric current flows through resistance wires and heats them up. This process is used, for example, in the plasticizing section of the cylinder.

roving According to German standard DIN 61850, rovings are a defined number of glass spun filaments (textile glass roving) combined in an approximately parallel arrangement to form a single strand. An individual glass strand in turn consists of a certain number of individual glass filaments which have been combined without twisting to form a thread of uniform yarn count in a mostly parallel order.

sandwich A planar multilayer composite design consisting of two high-strength outer layers and a lightweight thick inner layer. This design provides a higher surface moment of inertia and a high bending stiffness, respectively.

sealing point Time at which the molding compound in the sprue has solidified to such an extent that flow no longer occurs.

secondary valence forces	These are intermolecular forces which have a limited range of a few nanometers (= one millionth of a millimeter), e.g., hydrogen bonds.
self-extinguishing	Ability of a burning plastic to extinguish without external input of energy.
semi-finished product	A semi-finished product is an intermediate product made of plastic, for example tubes, films, and sheets, which are still further processed (formed) or finished into a final product.
shear	During plasticizing and injection, the molding material is moved at different speeds, i.e., the individual particles rub against each other. This process during injection molding is called shear.
shear heat	Shear heat is generated by converting mechanical energy into heat. This conversion takes place in the plasticizing cylinder, e.g., through friction between the pellets, on the screw and on the cylinder wall, during crushing and squeezing of the pellets, and finally through internal friction in the melt during plasticizing and injection.
shear modulus	The shear modulus refers to the shear strength of a material. Shear strength is a material parameter that describes the resistance of a material to shear, i.e., to separation by forces that seek to move two adjacent surfaces longitudinally.
shear rate	The shear rate or shear velocity defines the difference in flow velocity between two fluid layers and is expressed in units of 1/s.
shear strength (interlaminar)	Property defined as the ratio of the force leading to rupture failure within the area subjected to shear and the shear area (according to DIN 65148).
shear stress	Shear stress is the ratio of force to plate area during shear flow and has the unit Pascal (Pa).
shear-thinning fluid	Liquid whose viscosity depends on the shear rate.

sheet molding compound	Sheet Molding Compound (SMC) is a fiber composite technology used to produce long fiber reinforced molded parts. This process is used in large-scale production. However, it is not economical for small quantities.
shrinkage	Refers to a reduction in volume (volume contraction). Usual values are 0.2 to 2%, depending on the plastic.
small and medium enterprise	Small and medium-sized enterprises (SMEs) are companies of a defined size. Under EU guidelines (eligibility criteria for funding), SMEs may have a maximum of up to 249 employees (or up to €50 million in sales per year). Economic institutes set the upper limit at 500 employees.
softening temperature range	Within the softening temperature (ST) range, which is also referred to as the glass transition temperature (T_g) range (see there), the amorphous parts of a thermoplastic melt.
sonotrode	This is the welding tool used in ultrasonic welding. The sonotrode transmits the vibrations to the plastics part being welded.
stabilizers	Chemical additives that make a plastic more resistant to certain influences, such as UV radiation, heat, oxidation, weathering.
stagnation zone	Areas or locations in molds or other melt-bearing parts where the flow velocity is significantly reduced or even reduced to zero.
standard deviation	This is a measure of how much the individual values of a sample (for example, a series of measurements) deviate from the mean value.
state of matter	Plastics only have two states of matter: solid and liquid. Plastics decompose before they reach the gaseous state.
statistical process control	Statistical process control (SPC) is the statistical collection of process data to assess the quality of a product.

statistics	Scientific method for the numerical recording of stock and size changes and for the calculation of probabilities.
strain (at maximum load)	Strain is the change in length experienced by a body pulled in one direction under the action of a force. The strain at maximum force is that experienced by the body at the maximum force. It is expressed as a percentage of the initial position.
substitute fuels	Substitute fuels or refuse-derived fuels (RDF) belong to the class of secondary raw materials that are generated in a recycling process. These are, for example, plastics, paper, wood or treated sewage sludge that are produced as reusable raw materials during the sorting of waste.
synthesis	Creation of chemical compounds from the elements or more simply constructed basic chemicals to form new substances (Greek: synthesis = to put together).
temperature control system	The temperature control system (responsible for the cylinder temperature) is divided into several zones that can be heated or cooled independently from one another.
tensile strength	The tensile strength $\sigma_M = F_{max}/s_0$ is the stress resulting from the maximum force relative to the initial cross-section s_0.
thermal conduction	Thermal (also: heat) conduction, along with convection or heat flow and thermal radiation (see there), is one of the three mechanisms for transferring heat from one place to another. By means of conduction, heat is transferred through bodies from areas of higher temperature to areas of lower temperature. The thermal conductivity of different materials varies. Plastics are poor conductors of heat and therefore make good insulating materials.

thermal radiation	Thermal radiation, along with convection or heat flow and thermal conduction (see there), is one of the three mechanisms for transferring heat from one place to another. In thermal radiation, heat is transferred by electromagnetic waves (infrared radiation).
thermoplastic	Thermoplastics are non-crosslinked plastics that are meltable and can be solved. This group of plastics is divided into amorphous and semi-crystalline thermoplastics. The latter have crystalline and amorphous regions.
thermosets (duromers)	Polymers in which the molecular chains are linked three-dimensionally by covalent bonds. Often also referred to as duromers. They can no longer be deformed after curing.
three-zone screw	One of the most common screw shapes. It consists of feed zone, compression zone and metering zone.
unidirectional	Unidirectional means oriented in one direction.
variation range	Random variation range, extrapolated range from several samples of a characteristic, in which all further measured values will lie with a high degree of probability.
viscoelastic	Refers to the condition of a body that is both elastic (Hookean body) and viscous (Newtonian body).
viscosity	A measure of the flow resistance of a liquid. The higher the viscosity, the higher the flow resistance and the more viscous is the liquid.
viscosity curve	Shows the dependence of the viscosity on the shear rate of a fluid in a double-logarithmic plot.
vulcanization	Vulcanization is a chemical cross-linking process that transforms natural rubber into vulcanized rubber and enables its dimensional stability and elasticity.
wall adhesion	When a fluid flows alongside a wall, the flow velocity of the particles closest to the wall is zero; they adhere to the wall.

waste fraction

Waste fraction refers to the respective portions of plastic waste that can be processed as a whole. In chemistry, fraction (Latin: fractio = fraction) refers to a subgroup of substances in a mixture of substances, irrespective of their state of matter.

waste incineration plant

A waste incineration plant (WIC) is used for thermal waste treatment or recycling (in Switzerland: "Kehrichtverbrennung" or "Kehrichtverwertung"). It involves the incineration of the combustible fractions of waste for the purpose of reducing the volume of the waste using the energy it contains. This takes place in special incineration plants.

waste management

Waste management is the term used to describe all activities and tasks relating to the avoidance, reduction, recycling, and disposal of waste. This includes all waste from industry, trade, and the service sector as well as waste from households and public areas such as parks and streets.

water-assisted injection molding

Water injection technology (WIT) is a special injection molding process by injecting water into the plastic to produce hollow-walled bodies.

wettability

The wettability is a basic prerequisite for a successful adhesive bond when bonding plastics. The wettability is described by the surface energy. The wettability of plastics can be assessed quickly and easily by applying some water to the surface. If a water drop forms, the surface is of low energy. If the water runs, the surface is of high energy.

Appendix
Answers

Answers to Performance Review Questions

Lesson 1
1. thermosets
2. semicrystalline
3. meltable
4. insoluble
5. loosely
6. non-meltable
7. lighter
8. lower
9. different
10. good
11. can be
12. polycarbonate (PC)

Lesson 2
1. natural gas
2. cracking
3. propylene
4. chain
5. polymer
6. carbon (C)
7. poly
8. tangled
9. carbon (C)
10. optical
11. double bond
12. "coupling"
13. polypropylene (PP)
14. "separation"
15. "exchange"

Lesson 3

1. atomic bonds
2. intermolecular bonds
3. stronger
4. semicrystalline
5. crystal clear
6. strongly
7. swellable
8. transparent
9. primary shaping
10. reshaping
11. bonding
12. sawing
13. reshaped
14. welding

Lesson 4

1. elongation at break
2. strength
3. toughness
4. below
5. amorphous
6. rigidity
7. cross-linked
8. +130
9. can be

Lesson 5

1. strength
2. 1000
3. depends
4. creep
5. heated up
6. orientation
7. time and temperature
8. time-dependent creep diagrams
9. more

Lesson 6

1. lighter
2. 0.9 to 2.3
3. 2000
4. metal powder
5. roughly equal to
6. light transmission
7. 41
8. 12

Answers

Lesson 7
1. surface
2. shear rate
3. is independent of
4. alignment
5. flow resistance
6. decrease
7. Newtonian
8. decrease
9. shear thinning
10. zero viscosity
11. 10

Lesson 8
1. tear strength
2. transparency
3. can
4. snap together
5. material costs
6. more
7. inferior
8. temperature resistant

Lesson 9
1. processing
2. mixers
3. weight
4. plasticizing
5. more freely than
6. cutting mills

Lesson 10
1. continuously
2. the extruder
3. three-zone-screw
4. high
5. shape
6. multiple layers
7. extrusion blow molding
8. biaxially stretched
9. extrusion or coextrusion
10. extrusion or coextrusion

Lesson 11

1. primary processing
2. commodities
3. cycle time
4. finished parts
5. mold
6. shrinks
7. cooling time
8. screw
9. 2.9
10. side by side

Lesson 12

1. matrix
2. 590
3. b) impregnation
   c) shaping
   d) curing
4. hand lay-up
5. a) thermosetting
   b) thermoplastic
6. thermosetting
7. sailboats and tennis rackets

Lesson 13

1. gas bubbles
2. lighter than
3. uniformly
4. higher
5. not the same
6. foaming methods
7. high pressure process
8. injection molding

Lesson 14

1. heated up
2. thermoplastics
3. infrared radiation
4. only one side
5. prestretched
6. shorter than
7. PS

Lesson 15

1. added
2. fast
3. generative manufacturing
4. stair-stepping
5. well bonded
6. support structures
7. build-up direction
8. commodity plastics

Lesson 16
1. thermoplastic
2. cannot
3. viscosity and melting temperature
4. heat conduction
5. Hot gas welding
6. welding shoe
7. laser transmission welding

Lesson 17
1. less
2. thermal expansion
3. smaller than
4. can
5. highest
6. liquid cooling
7. must

Lesson 18
1. can
2. cohesion and adhesion
3. clean
4. peeling
5. thermosets
6. quickly
7. bigger
8. Even high-quality

Lesson 19
1. 370
2. a) short-life
   b) long-life
   c) long-life
3. 45
4. a lot of
5. hardly
6. preferable

Lesson 20
1. possible
2. chemical
3. energy and raw materials
4. be recycled
5. shredded
6. cleanly
7. sorted by type
8. high
9. recyclable
10. valuable
11. important
12. 90